Springer Geography

The Springer Geography series seeks to publish a broad portfolio of scientific books, aiming at researchers, students, and everyone interested in geographical research. The series includes peer-reviewed monographs, edited volumes, text-books, and conference proceedings. It covers the entire research area of geography including, but not limited to, Economic Geography, Physical Geography, Quantitative Geography, and Regional/Urban Planning.

More information about this series at http://www.springer.com/series/10180

Wossenu Abtew · Shimelis Behailu Dessu

The Grand Ethiopian Renaissance Dam on the Blue Nile

 Springer

Wossenu Abtew
Water Resources Division
South Florida Water Management District
West Palm Beach, FL, USA

Shimelis Behailu Dessu
Department of Earth and Environment
Florida International University
Miami, FL, USA

ISSN 2194-315X   ISSN 2194-3168   (electronic)
Springer Geography
ISBN 978-3-030-07301-5   ISBN 978-3-319-97094-3   (eBook)
https://doi.org/10.1007/978-3-319-97094-3

This Springer imprint is published by the registered company Springer Nature Switzerland AG
The registered company address is: Gewerbestrasse 11, 6330 Cham, Switzerland

Preface

Global population growth is putting stress on land, water resources and the environment. Deforestation, soil loss, desertification and climate change compound the challenges in the Nile basin. Population growth and increase in food, water and energy demand and associated internal social pressures will inevitably lead to water conflicts within countries and between countries. Transboundary or international basins and rivers have potential for conflict between countries which share the resources. Transboundary water rights vary from basin to basin and water right is a function of power order. Indus River, India and Pakistan; Colorado River, United States and Mexico; Mekong River, China, Myanmar, Thailand, Laos, Cambodia and Vietnam; Parana River, Brazil, Paraguay and Argentina; Euphrates River, Turkey, Syria and Iraq; Ganges River, Nepal and India; and other transboundary river basins water rights are unique that developed through time. It should also be noticed that land and water conflicts within entities in a country are equally important. There are 261 transboundary basins covering 45% of the global land surface. About 146 countries share transboundary basins. The Nile basin lies within eleven countries with total population of over 450 million that will reach 700 million in less than 25 years. The Nile basin growing land and water demand is not limited to the population in the river basin. In the absence of water resources use and basin management agreement in the Nile basin, unilateral water control and use projects will continue to advance to mitigate food and power shortage and relieve social pressure in each riparian country. The water control projects include out of basin water transfer and may include water trading. Each country cites legal basis or treaties that support its water claims. The Grand Ethiopian Renaissance Dam (GERD) is a reflection of current stressors in the basin with far-reaching conflict potential within Ethiopia and with riparian countries. The current political system and constitution of Ethiopia are based on regional ethnic federal structure with constitutional right to secede. This arrangement creates internal water right issues as currently observed inter-ethnic conflicts on land and borders. The GERD has potential for ethnic conflict from inequitable sharing of benefits from trans-ethnic waters, the construction economy of the dam, its operation and associated economic outputs.

Egypt's concern on potential flow reduction is demonstrated throughout the dam-related dealings between Ethiopia, Egypt and Sudan. Sudan appears indifferent or supporting the dam as it will benefit in many ways. Will there be enough water for all? The three countries have agreed for two French companies to undertake hydraulic and environmental studies to forecast the impact of the dam on downstream. The result of the study has the potential to ignite the conflict between Ethiopia and Egypt. One of Egypt's concerns is the number of years of initial filling of the dam as it will be a time of historical flow reduction unless the filling years are wet years. The longer the filling years, the lesser the flow reduction will be but the economic value of the dam will diminish. The dam is already overdue, and longer filling period could make it economic loss with all factors considered. Drought condition during filling will exasperate disagreements. Optimal power generation of the dam is questionable with unresolved upstream and downstream water right issues.

In this book, the hydrology of the Blue Nile basin is presented. The Nile River transboundary water rights; land and water rights in the Blue Nile basin, in Ethiopia; the GERD site and Ethiopian internal condition; GERD design analysis; GERD initial dam filling; dam operations for hydropower and upstream and downstream water rights; dialogue and diplomacy through GERD construction; finance sources of GERD; and aquatic weed potential on GERD reservoir are covered. This book is beneficial for students, academics, sociologists, engineers, policy-makers, water resources and environment professionals, the people of the Nile basin and everyone with interest on global land and water stress, population growth and water conflict.

West Palm Beach, USA   Wossenu Abtew
Miami, USA   Shimelis Behailu Dessu

Contents

Abbreviations

bcm	Billion cubic meter
BNB	Blue Nile Basin
CIA	Central Intelligence Agency
EPRDF	Ethiopian People's Revolutionary Democratic Front
ESA	European Space Agency
ENSO	El Niño Southern Oscillation
FAO	Food and Agriculture Organization
GCM	Global Circulation Model
GERD	Grand Ethiopian Renaissance Dam
GSE	Geology Society of Ethiopia
ha	Hectar
HAD	High Aswan Dam
HP	Hydropower
I	Irrigation
IFT	Illegal Fund Transfer
IPRC	International Pacific Research Center
m asl	Meters above sea level
MoWR	Ministry of Water Resources
NBI	Nile Basin Initiative
RCC	Roller Compacted Concrete
SGCC	State Grid Corporation of China
SOI	Southern Oscillation Index
SRTM	Shuttle Radar Topography Mission
SST	Sea Surface Temperature
TPLF	Tigray Peoples Liberation Front
USBR	United States Bureau of Reclamation
WUA	Water Use Association

Chapter 1
Introduction

Abstract The Nile Basin is one of the largest basins in the world shared by eleven countries. The principal tributaries of the Nile River are the White Nile, flowing from the Great Lakes region of Central Africa and the Blue Nile (Abay), Sobat (Baro-Akobo) and the Atbara (Tekeze), flowing from the highlands of Ethiopia. Ethiopia contributes close to 85% of the Nile river flow. The Nile basin is entering into a new era of challenges and opportunities primarily driven by population explosion, food and water shortage, increase in water demand and water use, climate change, and complicated water right issues. More importantly, upstream countries started to assert their right to develop the Nile water resources challenging the long-held water right hegemony of Egypt and Sudan. Ethiopia unilaterally launched the construction of Grand Ethiopian Renaissance Dam (GERD). The sheer size and storage capacity of GERD has initiated dialogue and diplomacy towards understanding of the current reality in the basin as well as the absolute need of co-operative water resource development. This chapter provides an overview to the Nile basin along with the social, economic, environmental and political implication of GERD. The book mainly focuses on the Blue Nile basin, the GERD design, filling and operation in association with the larger Nile basin.

Keywords The Nile · Blue Nile · Grand Ethiopian Renaissance Dam
Ethiopia · Egypt · Sudan · Transboundary rivers

1.1 Overview of the Nile River Basin

As a result of geological processes, rivers cross political boundaries creating dependence for diverse societies with a chance for cooperation and potential for conflicts. The history of people and their relation to rivers transcends a mere dependence for livelihood. The Nile River flows from the wet equatorial lakes region and high lands of Ethiopia to the dry desert regions of North-East Africa creating historical dependence for survival in Egypt and Sudan and support the livelihood and rich ecosystem and society of the downstream countries. The River Nile is the longest river in the

© Springer International Publishing AG, part of Springer Nature 2019
W. Abtew and S. B. Dessu, *The Grand Ethiopian Renaissance Dam
on the Blue Nile*, Springer Geography, https://doi.org/10.1007/978-3-319-97094-3_1

world stretching nearly 6700 km, covering more than 35° of latitude and draining an area of over 3 million square kilometers–one tenth of Africa's total land mass. The Nile River Basin traverses varied landscapes, with high mountains, tropical forests, woodlands, lakes, savannas, wetlands, arid lands, and deserts, culminating in an enormous delta on the Mediterranean Sea. For millennia, this unique waterway has nourished varied livelihoods, an array of ecosystems, and a rich diversity of cultures. The principal tributaries of the Nile River are the White Nile, which begins in the Great Lakes region of Central Africa; and the Blue Nile (Abay), Tekeze (Atbara), and Baro-Akobo (Sobat) flowing from the highlands of Ethiopia. The Sobat flows from southern Ethiopia to join the White Nile. The other sources are the equatorial lakes and the Bahir el Ghazal basin (Fig. 1.1). The most distant source is the Kagera River, which winds its way through Burundi, Rwanda, Tanzania, and Uganda into Lake Victoria. Eighty-five percent of the water of the Nile originate from the Ethiopian plateau and 15% for the Equatorial Plateau (Moges and Gebremichael 2014).

The Nile River is shared by 11 countries: Burundi, Democratic Republic of Congo, Egypt, Eritrea, Ethiopia, Kenya, Rwanda, South Sudan, Sudan, Tanzania, and Uganda with varying flows with time along its path (Fig. 1.2). It is home to world-class environmental assets, such as Lake Victoria (the second-largest fresh water body by area in the world) and the vast wetlands of the Sudd. It also serves as home to an estimated 160 million people within the boundaries of the basin, while about twice that number, about 443 million-live in the 11 countries that share and depend on Nile waters projected to increase to 726 million in 20 years (Abtew and Melesse 2014a).

Upstream wetter riparian countries were not actively seeking to share water until population growth started putting pressure to utilize Nile resources and meet their growing water demand. At the same time, the development and access to technological knowhow on water control such as pumps, dams, canals, water harvesting and irrigated agriculture is increasing upstream water use. The construction of GERD is a reflection of the slow and often discouraging co-operation among riparian countries and the contemporary economic development and socio-political dynamics creating the need for an increasingly aggressive infrastructure development. The future of the Nile is increasingly scarce resource with a large population and infrastructure that depends on the Nile water. An open and transparent dialogue and co-operations are needed more than ever to set the right course towards equitable sharing of water and associated benefits before the fragile regional hydro-politics evolve into water conflict. GERD is about half the storage volume of the High Aswan Dam well above the annual yield of the Blue Nile River. Coordination of all the dams in series could lessen the impact of GERD filling for a shared common goal to reduce undesirable consequences and promote regional accord. The move by Egypt and Sudan to walk out of the Nile Basin Cooperative framework signed by the upstream countries may hinder such cooperation that would have set a precedence in future infrastructure development. The race for water right assertion through unilateral projects is likely to continue. Hence, initial filling and operation of the GERD impact could be minimized through synchronized dam operations along the Nile.

By 2025, 1.6 billion people in the world will live under water scarcity (Eliasson 2015). Water conservation and watershed management, basin planning and basin

Fig. 1.1 The Nile River Basin and major dams

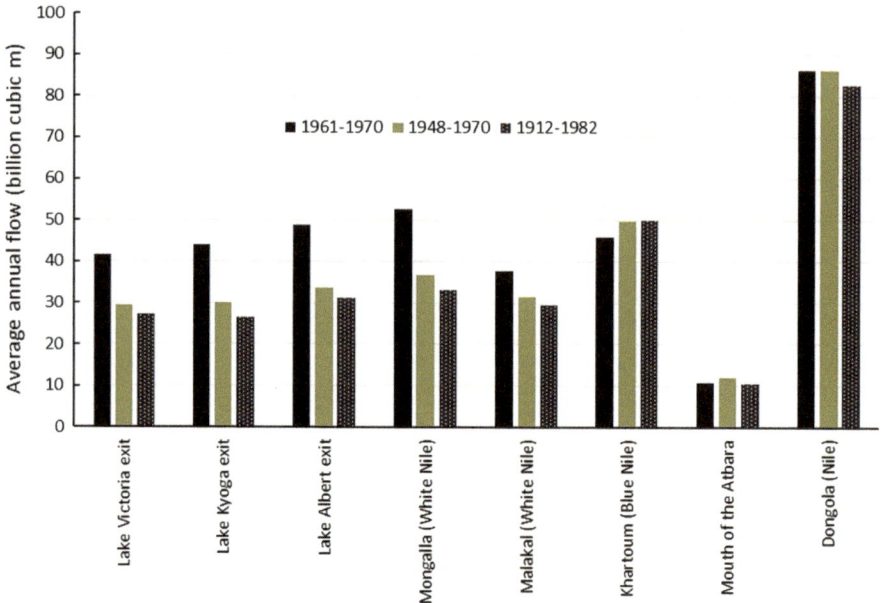

Fig. 1.2 Averaged annual flow (billion cubic meters, bcm) at selected stations on Nile River (Abtew and Melesse 2014a, b; original data source Karyabwite 2000)

wide cooperation would make the most use out of the water and minimize conflicts. Water transfer outside the basin may expand. To reduce vulnerability to climate change water cooperation for the Nile basin is advised (Hammond 2013).

1.2   Water Resources of the Nile Basin

The Nile Basin comprises of countries with an average annual rainfall of more than 1300 mm and others almost none throughout a given year. The economic status of the riparian countries also varies from a capacity that enables an almost full utilization of available water resource to countries with no significant project implementation in their part of the Nile basin.

Water resource potential of a country in the Nile basin could be viewed, assessed and reported from different viewpoints. For instance, 98% of Burundi is in the Nile basin while the major contribution in terms of flow quantity comes from the highlands of Ethiopia where about 32% of the land is in the Nile basin. Hence, the water resource potential of the riparian countries has an absolute as well as relative expression in the context of the respective nations and the Nile basin as a whole.

1.3 GERD and the Blue Nile Basin

Even though Nile Basin is shared by eleven countries; Sudan and Egypt highly depend on the flow from the Ethiopian Blue Nile with contribution up to 60% of the Nile flow while covering only 10% of the Nile basin drainage area. The Blue Nile River starts from the Highlands of Ethiopia and flows to Sudan joining the White Nile River in Khartoum (Fig. 1.3). GERD is the first major dam on the Blue Nile (Abay) River of Ethiopia, whereas Sudan has the Rosaries and Sennar dams on the Blue Nile.

GERD is a combination of 1.8 km long high gravity dam, 5 km long rockfill saddle dam, and a 300 m long separate spillway between the main and saddle dam, on the Blue Nile (Abay) River of Ethiopia. It is about 20 km from Ethio-Sudan boarder (Fig. 1.4).

1.4 The GERD, Ethiopia, Egypt and Sudan

The promising cooperative effort among riparian countries with the launch of the Nile Basin Initiative in 1999 fall short of its goal when co-operative framework agreement was signed by upstream countries with strong opposition from Egypt (Hammond 2013). The fallout from the negotiation of Eastern Nile riparian countries (Ethiopia, Sudan, and Egypt) on the commissioning of cascade dams in Ethiopia was a major driver for the Ethiopian government to take a unilateral step towards the construction of GERD. The proposed dams were Kara Dobi, Mabil, Mendia and Border. GERD is located within a few kilo meters from the most downstream Border site close to the border with Sudan. Afterwards, the Ethiopian government has surprised the Nile riparian countries with the launch of GERD construction in 2011. The planning and design of the dam was kept secret and the construction was portrayed as a national land mark to symbolize economic success, and national security and to some extent score political advantages bringing internal unity through nationalism. However, the GERD has also been marred with domestic challenges from political oppositions and fallouts of international and regional diplomacy particularly in securing funds.

Ethiopia has repeatedly attempted to give assurance that downstream flow will not be reduced by GERD. Ethiopia argued the choice of the GERD site as a token of consideration to downstream countries where the dam can only be used to generate hydroelectric power with practically no consumptive withdrawal of water for irrigation. However, the dialogue has shifted over the last five years from Ethiopia's considerations on site selection to the size of the dam and the potential impact during the initial filling of the dam. The Ethiopian government argues the filling will have minimal impact rather benefits downstream riparian countries with regards to flooding, siltation, irrigation and water conservation (http://www.mowie.gov.et/dams-and-hydropower accessed on 11/12/2017). These benefits and Ethiopia's right to develop its water resources for the development and prosperity of its people has been

6 1 Introduction

Fig. 1.3 The Blue Nile basin of Ethiopia and Sudan

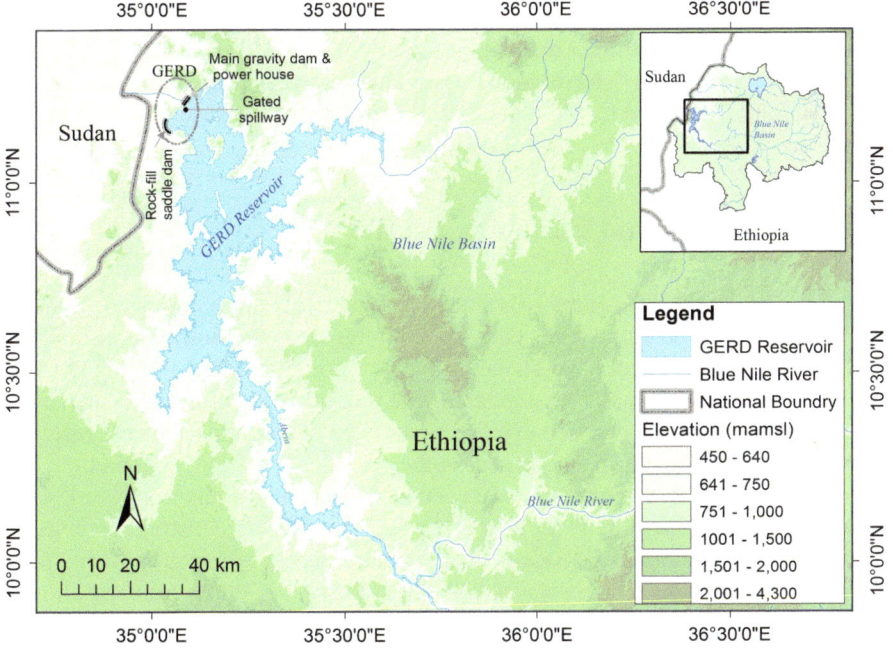

Fig. 1.4   Location of GERD and the reservoir extent at full supply level

rejected by Egypt (Whittington et al. 2014). Egypt has been engaged in maintaining the status-quo despite the reality of GERD (Tawfik 2016). Kahsay et al. (2015) showed that the negative effects of GERD on Egypt's economy will reverse once the dam become fully operational providing considerable benefit to both Sudan and Egypt in the long run. (Gebreluel 2014) argued that Egypt's lack of interest to focus on potential future cooperation is due to the potential of GERD to disrupt Egypt's historical monopoly on the Nile waters. Ethiopia, Sudan and Egypt signed the Declaration of Principles propelled by the reality of GERD despite alleged effort by Egypt to deter the progress of construction (Salman 2016).

Egyptians have water resources expertise to model the Blue Nile basin hydrology and GERD hydraulics and evaluate potential downstream impact. Tensions have been increasing between Ethiopia and Egypt in recent years with Ethiopia accusing Egypt for having hand in uprisings in Ethiopia. Ethiopia, Sudan and Egypt signed an agreement in October 2016 for two French consultancy firms, BRL and Artelia, to perform a water resources study and hydropower simulation modelling and downstream environmental and social impact of the dam. When the technical report comes out with potential downstream impact, Ethiopia may run out of explanations as its position was not founded on water rights. It may pursue on its political wrangling with Egypt skirting technical issues concerning the dam.

The economic viability of the GERD is founded on quality of construction, cost management, construction time, initial dam filling period, upstream water demand, downstream water demand, optimal reservoir operation for power generation with limited constraints, sediment deposition rate, dam and power line security, power price, market availability and internal and regional peace. Nile basin cooperation and water sharing framework could reduce conflict and make best use of all resources.

1.5 The GERD and Ethiopian Politics

Ethiopia contributes 85% of the Nile waters. The political and economic significance of Nile waters from Ethiopia has been linked to regional and domestic instabilities. Previous Ethiopian rulers have blamed Egypt for organizing and funding domestic uprising including the cessation of Eritrea. The recent history of Ethiopia and its politics is a reflection of population explosion, shortage of essentials and as a result it is plagued with resource grabbing based on the primal association, constitutionalized ethnicity. Every urban and rural project can be defined under this framework with a group or groups benefiting from land and water acquisition, construction contracts acquisition, employment, displacement and replacement of ethnic groups, and securing projected benefits from projects. The GERD is one of these projects. The dam will be detrimental to the indigenous people of the reservoir site and surrounding areas but likely benefits businesses and industries that arise around the dam. The environment is set for displacement and replacement of ethnic groups. There is indications of this expressed in resettlement related conflicts in the Benishangul region. There is also a concern and active citizen follow-up that water rights of upstream people can be given away through transboundary agreements without the people's awareness.

The interest of multinational companies to acquire agricultural land in the Nile basin has brought out of basin interests into the Nile basin land and water arena that could influence water policy and agreements (Abtew 2014). The general global food shortage threat has initiated countries and businesses with capital to acquire land and water resources of other countries and transport the product back home. A very good account of land acquisition and water is presented by Fred Pearce (2012). The Gambela region in Ethiopia is at the center of land acquisition in the Nile basin in Ethiopia where indigenous people were reportedly affected (UNICEF/Addis Ababa 2006; Oakland Institute 2015).

1.6 The Way Forward with GERD

Unless extraordinary natural and political circumstances halt the construction, GERD is going to be the new reality in the Nile Basin. The immediate impact will be reduced flow to Sudan and Egypt. The recession agriculture on the banks of the Blue Nile will likely experience the immediate impact of GERD. Even though these immediate impacts will gradually subside, the recession agriculture practices will need to adapt to the advantages of regulated uniform water delivery due to the attenuation of rainy season flood flows by GERD. Other economic activities that has been relying on inundation of the Blue Nile banks such as brick production will also be affected. However, Sudan will benefit from reduced silt flowing to the Rosaries and other reservoirs along the Blue Nile and Main Nile Rivers.

Watershed management of the Blue Nile watershed has been neglected by upstream and downstream countries. With increasing water needs and limited water supply, basin management and cooperation need will surface. There is a growing water utilization in the upper Blue Nile Basin where farmers are employing small scale water harvesting practices. There is a push from local governments in Ethiopia to secure food production to the growing population in the Blue Nile Basin. The sum of these individual and local water utilization will ultimately reduce the Blue Nile flow unless the downstream countries actively work with Ethiopia in improving agricultural practices and assist in watershed management.

Egypt and Sudan may need to acknowledge the long-term advantage of GERD and engage in constructive dialogue and active participation on the successful completion and operation of the dam. GERD reservoir has a much lower evaporation loss than any of the downstream storages in Sudan and Egypt improving overall water storage efficiency. The dam also provides the much needed regulated flow to the thriving irrigation on the fertile lands of Sudan with minimal infrastructure investment.

1.7   Summary

The Nile River drains and flourishes a diverse culture and ecosystem of eleven countries from East-central Equatorial region, from Ethiopian plateau to Northern arid regions of Egypt. The basin has geo-political strategic importance and the water resource utilization in the basin has a complex regional and global implications. Since the 1970s, the High Aswan Dam has been a symbol of Egyptian prosperity and regional hegemony on the Nile water. However, after a series of diplomatic fallout to initiate co-operative physical water infrastructure to promote sustainable development of the upstream countries, Ethiopia launched construction of the Grand Ethiopian Renaissance Dam (GERD) on the main Blue Nile (Abay) River close to Ethio-Sudan border. The construction of GERD has since resurfaced the complex internal and regional issues associated with development and utilization of the Nile waters.

GERD is setting precedent in the basin in social, political, economic and environmental dimension. The dam has elevated hydro-politics awareness and helped to fuel national interests in the riparian countries. Ethiopia's self-reliance to fund the dam may embolden other upstream countries to take similar approach and commission projects of varying scale. National funding and Chinese involvement in infrastructure development of sub-Saharan counties may balance western financial institutions lack of support for upstream Nile projects.

There has been considerable concern on the downstream environmental impact of GERD. However, the environmental degradation and change in land management in the upper Blue Nile River is not appreciated enough and is likely to have a ripple effect in reducing the flow of Nile irrespective of GERD. Watershed management in the Nile basin is the interest of both upstream and downstream users.

Despite Ethiopia's assertion of 'no harm' to downstream countries, Egypt has been channelling concerns that evolved from the lack of transparency on the planning and design of the dam to the structural and operational safety of the dam to initial filling scenarios and reservoir size. Multiple diplomatic efforts to address these issues have been going with a slow pace and apparently no tangible agreement beyond common understanding. The size of GERD will cause temporary constraints and water shortage on the downstream countries. It is of the utmost importance that the initial filling of GERD be coordinated with downstream reservoirs to reduce these undesirable consequences. However, the three countries must look at the long-lasting shared benefits from GERD beyond the temporary challenges of filling the dam. The challenges and opportunities presented by GERD transcend the geo-political boundaries of the riparian countries.

The impact of climate change is yet known with certainty. Desertification expansion from northern Africa southward is documented (Abtew and Melesse 2016). Will there be enough water for everyone in the Nile basin is a scary question?

References

Abtew W (2014) Chapter 7: land and water in the Nile basin. In: Melesse AM, Abtew W, Setegn SG (eds) Nile River basin ecohydrological challenges, climate change and hydropolitics. Springer, New York

Abtew W, Melesse AM (2014a) Chapter 2: The Nile River basin. In: Melesse AM, Abtew W, Setegn SG (eds) Nile River basin ecohydrological challenges, climate change and hydropolitics. Springer, New York

Abtew W, Melesse AM (2014b) Ch. 28 Transboundary rives and the Nile. In: Melesse AM, Abtew W, Setegn SG. Nile River Basin ecohydrological challenges, climate and hydropolitics. Springer, New York

Abtew W, Melesse AM (2016) Landscape dynamics and Evapotranspiration. In: Proceedings of the world environmental & water resources congress, ASCE 22–26 May 2016

Eliasson J (2015) The rising pressure of global water shortages. Nature 517: 1 Jan 2015

Gabreluel G (2014) Ethiopia's Grand Renaissance Dam: ending Africa's oldest geopolitical rivalry? Wash Q 37(2):25–37

Hammond M (2013) The Grand Renaissance Dam and the Blue Nile: implications for transboundary water governance. Global Water Forum, Canberra, Australia

Karyabwite DR (2000) Water sharing in the Nile River valley. UNEP/DEWA/Grid, Geneva

Khasay TN, Kuik O, Brouwer R, van dar Zaag P (2015) Estimation of the transboundary economic impacts of the Grand Ethiopian Renaissance Dam. A computable general equilibrium analysis. Water Resources and Economica 10(Supplement C):14–30

Moges SA, Gebremichael M (2014) Chapter 18 Climate change impacts and development-based adaptation pathway to the Nile River basin. In: Melesse AM, Abtew W, Setegn SG (eds) Nile River basin ecohydrological challenges, climate change and hydropolitics. Springer, New York

Pearce F (2012) The land grabbers: the new fight over who owns the earth. Beacon Press, Boston, MA investment

Salman SMA (2016) The Grand Ethiopian Renaissance Dam: the road to the declaration of principles and the Khartoum document. Water Int 41(4):512–527

Tawfik R (2016) The Grand Ethiopian Renaissance Dam: a benefit-sharing project in the eastern Nile? Water Int 41(4):574–592

The Oakland Institute (2015) We say the land is not yours: breaking the silence against displacement in Ethiopia. April 2015, Oakland, California

UNICEF/Addis Ababa (2006) Livelihood & vulnerabilities study Gambela region. UNICEF/Addis Ababa, Ethiopia

Whittington D, Waterbury J, Jeuland M (2014) The Grand Renaissance Dam and prospects for cooperation in the eastern Nile. Water Policy 16(4):595–608

Chapter 2
The Nile River and Transboundary Water Rights

Abstract The Nile River is a transboundary river shared by 11 countries: Burundi, Egypt, Democratic Republic of the Congo, Eritrea, Ethiopia, Kenya, Rwanda, South Sudan, Sudan, Tanzania and Uganda. Among the riparian countries, Ethiopia is the largest contributor to the Nile River accounting to 82% of the annual flow. Nile River water rights are complex as a result of the basins geography climate and political history. Pre-colonial, colonial and post-colonial claims and assertions of water rights are being challenged due to growing population, awareness and water demand. The Blue Nile is the largest of the tributaries, a collection of tributaries in the Ethiopian Highlands, flowing through settlements, farm fields and natural landscapes. Water demand in the basin is continuously growing as a result of population growth, climatic factors, and small to medium scale irrigation and water harvesting schemes. Small dams and control structures are being built on tributary rivers for both hydropower and irrigation. Ethiopia has been investing on multiple dams on Nile river tributaries and is building a major dam, the Grand Ethiopian Renaissance Dam (GERD) on the Blue Nile since 2011. Although, there are over thirty dams and water control structures on the Nile and tributaries, dams on the Blue Nile raise attention as it is the main source of water for the Nile, The GERD has brought to the forefront the question of water rights of the Nile with a tug of war between Ethiopia and Egypt with the main concern of downstream flow reduction as the result of the dam. Negotiations and agreements have been murky as water rights and water sharing is not addressed directly. This chapter outlines the history of water rights, water use and water controls in the Nile basin, and current realities associated with the construction and operation of the GERD.

Keywords Nile River · Blue Nile · Transboundary Rivers · Ethiopia Egypt · Sudan · Water rights · Ethiopian Renaissance Dam

© Springer International Publishing AG, part of Springer Nature 2019   13
W. Abtew and S. B. Dessu, *The Grand Ethiopian Renaissance Dam
on the Blue Nile*, Springer Geography, https://doi.org/10.1007/978-3-319-97094-3_2

2.1 Overview

The Nile River is the longest river in the world, 6853 km, originating from or crossing through eleven countries from east and east-central Africa to the Mediterranean Sea (Fig. 2.1). The drainage area is about 3.17 million km^2 (FAO 2007). Even though Nile is the longest river compared to major rivers basins, the flow volume is not large compared to major rivers. It is 99th in flow rate compared to 151 major rivers on all continents. The average flow rate is 2810 m^3 s^{-1}, 89 bcm per year (Wohle 2007). The Nile river system can be identified in four regimes; water source, water accumulation (energy source), water losing and water consuming (Moges and Gebremichel 2014). These regimes are identifiable from source to consumption at the terminal. Estimated historical annual average flow of the Nile at Aswan is 84.1 billion m^3 (bcm), Sutcliffe and Parks (1999). The major sources are the Blue Nile (Abay), Atbara and Sobat from Ethiopia and the White Nile from drainage areas in Burundi, Rwanda, Kenya, Democratic Republic of the Congo, Tanzania, Uganda, South Sudan and Sudan. In general, the sources of water can be regionalized as the Ethiopian plateaus the Equatorial Lake Region and the Bahr El Ghazal Basin (Fig. 2.1). Ethiopia contributes 82% of the Nile flow through the Blue Nile, Atbara and Sobat. Blue Nile (Abay) basin constitutes about 40% (55 bcm) of the total water resource of Ethiopia and is one of the least utilized basin.

   GERD is the first major Dam on the Blue Nile River of Ethiopia. It is located near to the Ethio-Sudan Boarder. The Blue Nile River crosses the Roseires and Sennar Dams before joining the White Nile in Khartoum, Sudan, forming the Nile River. The High Aswan dam in Egypt is the largest operational dam so far in the Nile Basin.

2.2 The Nile and Water Rights

The history of Nile water rights can be divided into pre-colonial, colonial and post-colonial periods. In earlier times, water rights may be irrelevant as both upstream and downstream societies could use as much as they need with down streamers not knowing where the source is and up streamers not caring for the destination of the river. Climatic fluctuation of river flow would have been the main concern with flow fluctuation credited to super natural forces. Historical records show excess floods and droughts with famines driving political and societal impacts.

2.3 Pre-colonial Period

Regional climate created downstream dependence on Nile River water while upstream solely depended on rain-fed agriculture with small scale irrigation. But concern on control of the source of the Nile for water security goes as far back as

Fig. 2.1 Location map of major dams on the Nile River and its tributaries

700 years. Historical records of military and diplomatic efforts to control source of the Nile is documented by Degefu (2003), Abtew and Melesse (2014) and Stoa (2014). Detailed study on treaties agreements and related convention from 1815 to 2008 is presented by Abseno (2009). In 1321 A.D., Ethiopian Emperor AmdeTsion sent envoy to Cairo to Sultan Al-Nasir Muhammad Q-Alaum protesting the persecution of Coptic Christians and threatening to divert the Nile unless they get relief (Degefu 2003). In 1513, the priest king Prester John brought a Portuguese explorer, Alfonso d'Albuquerque to look into diverting the Nile to the Red Sea through a tunnel with no practical outcome (Carlson 2013).

2.3.1 Colonial Period

The colonial period is marked by British control of Egypt, 1882–1956; Sudan, 1889–1956; Uganda, 1894-1962; Kenya, 1895–1964 and Tanzania, 1919–1961. Belgium controlled Burundi, 1916–1962; Rwanda, 1922–1961 and DRC, 1908–1960. Colonial period brought the recognition for water value, right, security, harnessing and use, and long-term insight. Although cotton existed in Egypt during the Roman times (2nd century), considerable large-scale production started on the Nile delta during Muhammed Ali's time (1805–1848), (Schanz 1913). The development of cotton spinning machine in 1738 and the cotton gin in 1793 increased the value of cotton worldwide. Cotton, Egypt, Britain and the Nile became linked with implication to water right claims to the Nile.

The British did their best to secure the Nile water for Egypt and Sudan. To maintain the flow of the Nile to the cotton fields of colonial Sudan and Egypt, the British initiated a treaty with the Ethiopian emperor Menelik while dealing with Ethio-Sudan border matters. The Ethiopian version of the 1902 treaty as written in Article II of the Amharic (Ethiopian official language) version can be translated as "Emperor Menelik II, king of kings of Ethiopia, has agreed not do work across the Blue Nile river, Lake Tana and Sobat river, that could block flow to the White Nile or allow others to do work without prior agreement with the English government". Law review by Mundy (2015) states that the treaty is vague; and change in circumstances surrounding the application of a treaty such as the end of colonization has been a ground for rejection of colonial treaties. More importantly, the Amharic version translates as "flow from the three water bodies may not be completely stopped".

Appointment of various commissions by the British for developing Egypt and Sudan started as back as 1884. The British Commission proposed to introduce cotton to the Sudan with a plan for two dams, Sennar on the Blue Nile and Jebel Aulia on the White Nile (Degefu 2003). A 1919 British Commission was set up to resolve opposing views whether the Sennar dam in Sudan will have impact on the Low Aswan dam in Egypt (built 1898–1902). This was followed by the 1929 agreement (exchange of notes) between United Kingdom and Egyptian government giving the rights of water to Egypt including upstream veto power on Nile water use and water control operations still allowing water use in Sudan. A British weekly, Wonders of

World Engineering, published an article on March 30, 1937, "The Nile under Control" which glamorized the engineering work of the Sennar dam and reclamation of the desert in Anglo-Egyptian Sudan. It states that Egypt has the right for Nile water including the silt that fertilizes its soil. The source of the Nile "Abyssinia" was casually mentioned. It also stated that the people of Egypt have claimed to control the Sudan to secure water rights. The 1922–1935 Ethio-Britain relationship was built around the interest of Britain to construct a dam at the outlet of the Blue Nile at Lake Tana. Ethiopia's attempt to appease foreign powers and initiation of the project laid down the foundation for Italian invasion of Ethiopia in 1935 (McCann 1981). After Britain was frustrated with Italy and Ras Teferi of Ethiopia unable to get into agreement on the Blue Nile and get a dam built by an American company, it resorted to unilateral control to Nile waters. It was suggested from British foreign office that Britain should ask the League of Nations for mandate to control the Nile waters independent of territories (Tvedt 2004).

2.3.2 Post-colonial Period

The post-colonial period questioned and undermined the colonial era water rights culminating with the 1999 formation of the Nile Basin Initiative (NBI) that was signed by all riparian countries except Eritrea. The objective of the NBI was '*To achieve sustainable socio-economic development through the equitable utilization of, and benefit from, the common Nile Basin water resources.*' Through the NBI, the Cooperative Framework was developed for equitable share and economic benefit of the basin water resources. The Cooperative Framework agreement was signed in 2010 by upstream countries Burundi, Ethiopia, Kenya, Rwanda, Tanzania and Uganda. The Ethiopian parliament ratified the agreement in 2013. Egypt and Sudan did not support the agreement as maintaining the colonial era water rights is to their best interest. Upstream countries have become more assertive of their water rights as a result of increased need for power and food with population growth. But, the share of water use has stayed asymmetrical due to uneven economic, social, political, diplomatic and military developments among Nile countries and uneven dependence on Nile waters.

Current estimates are Ethiopia contributes 72 bcm and use 1 bcm while Egypt has no contribution and uses 55 bcm. Sudan and South Sudan's contribution are minimal and they use 18.5 bcm. The Central African Equatorial Lakes countries contribute 12 bcm and consume 1.7 bcm (Kalpakian 2015). According to Sudan's minister for water resources, irrigation and electricity, from Sudan's share, 6.5 bcm is granted for Egypt temporarily in a separate agreement. The minister stated that the main concern of Egypt is that it will lose Sudan's share as a result of GERD (Sudan Tribune 21 November 2017). This could be one of the thorny water issues between Sudan and Egypt. The Central African Equatorial Lakes countries contribute 12 bcm and consume 1.7 bcm (Kalpakian 2015). Assertion of water rights and planning projects in both upstream and downstream countries has political and economic reasons mostly driven by population pressure. The initiation of the Grand Ethiopian Renaissance Dam in 2011 is the extension of this development which is based on increasing power need and food security. The legality of Ethiopia's right to build the dam based on the Vienna Convention on the Law of Treaties is confirmed in a law review (Mundy 2015). Ayman Salama, Professor of international law and member of the Egyptian Council of Foreign Affairs (ECFA) stated that Egypt has no right to request Ethiopia interrupt the dam construction as the March 2015 Declaration of Principles didn't include such clause (AhramOnline 1 March 2016). Without comprehensive water agreements and watershed management, water abstraction will continue with Nile tributaries getting depleted first. The decision of Egypt and Sudan to withdraw from the 2010 Cooperative Framework Agreement may have undesirable consequences on the efficient utilization of the Nile water and promote unilateral development that ultimately initiate regional water conflict.

2.3.3   Population Growth and Water Interest

When Britain and Egypt agreed to build a dam on Lake Tana in 1935 provided Ethiopia's consent, the population of Egypt was estimated around 15.6 million (Collins 1994). In 1954, Egypt's population grew to about 23.7 million when Uganda wanted to build the Owens Fall Dam off Lake Victoria. Since both Egypt and Uganda were under Britain, Egypt was able to enforce a condition to control the operations of the dam from the site to guarantee its interests.

The 1950s was a period where the value of water and transboundary river issues came to the forefront. Recognizing this fact, the United Nations started its several attempts to codify and bring to order the rules of transboundary water use by upstream and downstream interests. On November 21, 1959 the UN passed resolution 1301 (XIV) when the assembly dealt with codifying non-navigational uses of transboundary rivers. But, 3 months ahead, on August 8, 1959, Egypt and Sudan signed exclusive rights to Nile waters excluding upstream source countries. In 1959 Egypt's population was about 27.2 million (Fig. 2.2). Details of historical water claims on the Nile are given in Degefu (2003);

There are 34 major dams and numerous water control structures built along the Nile River and its tributaries (Table 2.1; Fig. 2.1). Initiation of these dams and other water control structures were associated with specific needs that correspond to rising population and water demand. As a result of population growth and pressure from food and energy security, upstream countries interest to benefit from the Nile came forward resulting in the formation of the Nile Basin Initiative in 1999 and the signing of the Cooperative Framework Agreement in 2010. Ethiopia started building the GERD in 2011 when its population reached 90 million. Figure 2.2 depicts historical and projected combined populations of Ethiopia, Sudan and Egypt with major Nile water control events. Expansion of recent water use by downstream countries mirrors their population and water demand growth.

Table 2.1 Dams and water control structures on the Nile River (Tesemma 2009; Conniff et al. 2012)

Dam/water control structure	Built	Country	Purpose, irrigation (I); hydropower (HP in MW)	River
Delta Barrage	1833–1862	Egypt		Nile
Aswan Low Dam	1898–1902	Egypt	I (5.1 bcm)	Nile
Assuit Barrage	1903	Egypt	I	Nile
Nag-Hamady	1930	Egypt	I	
High Aswan Dam	1960–1970	Egypt	I (132 bcm) and HP (2100)	Nile
Al-Sallam Canal	1997	Egypt	transfer water to Sinai	Nile
Fincha	1973	Ethiopia	134	Blue Nile tributary
CharChara	2000	Ethiopia	HP (84)	
Koga	2008	Ethiopia	I; HP (80)	Blue Nile tributary
Tekeze	2009	Ethiopia	HP (300)	Atbara
Tana-Beles	2010	Ethiopia	I (Lake Tana) HP (460)	Blue Nile
GERD	Under construction	Ethiopia	HP (6000)	Blue Nile
Megech		Ethiopia	I	Inflow to Lake Tana
Chemoga Yeda	2013	Ethiopia	278	Blue Nile tributary
Gumera		Ethiopia	I	Inflow to Lake Tana
Ribb	Under construction	Ethiopia	I	Inflow to Lake Tana
Didessa	Under construction	Ethiopia	I	Blue Nile tributary
EwasoNgiro	2012	Kenya	HP (180)	
Sennar	1925	Sudan	Irrigation (0.8 bcm)	Blue Nile
Jebel Aulia	1937	Sudan	Summer irrigation for Egypt	Nile
Kash el Girba	1964	Sudan	I (1.3 bcm) HP (10)	Atbara
Roseires	1966	Sudan	I (7.4 bcm) HP (1800)	Blue Nile
Merowe	2009	Sudan	I 12.5 bcm HP (1250)	Nile

(continued)

Table 2.1   (continued)

Dam/water control structure	Built	Country	Purpose, irrigation (I); hydropower (HP in MW)	River
Dal	Planned	Sudan	HP (400)	Nile
Shereyk	Planned	Sudan	I HP (350)	Nile
Kajbar	In progress	Sudan	I HP (30)	Nile
Rumela Dam	2011	Sudan	I ? HP 135	Atbara
Burdana Dam	2011	Sudan	I ? HP 120	Setit
Jonglei Canal	1978 not completed	Sudan	drain the Sudd wetlands	White Nile
Rusumo I & II	2012	Rwanda	HP (60)	Kagera
Nyabarongo	2012	Rwanda	Hp (27)	Kagera
Owens Fall (Nalubaale)	1954	Uganda	HP (180)	Nile (Lake Victoria)
Kiira	2000	Uganda	I HP 2000	Nile tributary
Isimba	2015	Uganda	HP(87)	White Nile

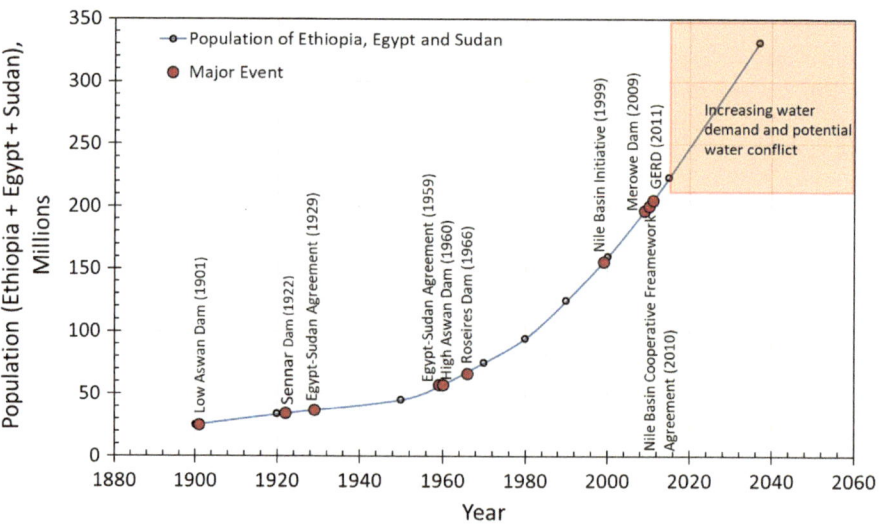

Fig. 2.2   Historical and predicted combined population growth in Egypt, Ethiopia and Sudan and significant events in the Nile River basin

2.3.4   River Diversion and Inter-basin Water Transfer

Current water use in the Nile basin can be summarized by constant increase in water use by each country, state and locality to set precedence of water use. Without water sharing agreement accepted by both upstream and downstream countries that includes watershed management and water conservation, abstractions in several forms will continue to be common with reduced water use efficiency. Various dams and water control structures have been built and several more are under construction and planned for future construction (Table 2.1). Increasing water need in the Nile basin is better illustrated by the Egyptian project that transfers Nile water to North Sinai across the Suez Canal through the El-Salaam Canal. The El-Salaam (Peace) Canal originates at Damietta branch and travels 144 km southeast crossing the Suez Canal and siphon through four tunnels (750 m long, 5.1 m diameter). The canal extends east by 175 km. According to Al Ahram (2 February 2016), the extension from the Suez Canal was completed in August 2015. It is planned to deliver 4.5 bcm where half of the water is directly from Nile River and the other half from Nile drainage (Othman et al. 2012). Egypt's New Valley (Toshka Canal) project is also a diversion project intended to convey Nile water to the western desert areas of Egypt.

Tanzania is planning a $27.6 million project to divert Lake Victoria to Kahama region in contradiction to the colonial period treaty between Britain, Egypt and Sudan (Warner 2011) which apparently Tanzania does not recognize. The ultimate scramble for water is river diversion, inter-basin water transfer, which reflects global water stress and may involve more Nile basin countries as water need keeps on growing. The next phase of mitigation to droughts and water shortage is adopting river diversions and inter-basin water transfer in water management plans. India and China are planning and pursuing inter-basin water transfer to mitigate drought and water shortages.

Ethiopia's surface water sources are runoff flows from the drainage basins. Unfortunately almost all the rivers, with the exception of Awash, are transboundary which may limit the range of utilization of the resources including the Nile. Forty four percent of the surface water resource in Ethiopia is in the Blue Nile basin (Fig. 2.3). Due to the current economic development and population growth, Ethiopia has no choice but develop and use the transboundary rivers in its watersheds despite the daunting diplomatic progress on equitable share of the resource with the downstream countries. Large scale land investment is currently estimated at 689,983 ha in the Nile basin in Ethiopia. Out of this 385,383 ha is in Pibor-Akobo-Sobat with proposed consumptive water use. El-Fadel et al. (2003) studied the Nile water issue and suggested a way forward could be addressing the Nile water issue by an international committee that could develop strategies for water conflict resolution and harmonious use of the resource. A detailed account of water scarcity, food security, population growth along the line and the prospect of water conflict is documented by Oestigaard (2012).

2.4 Dams on the Nile

Water control on the Nile goes as far back as 2650 BC when Sadd el-Kafara, a masonry dam, was built to control flooding in Egypt but failed due to high flows leaving its mark as the oldest dam of its size (Mays 2010). The earliest major dam is

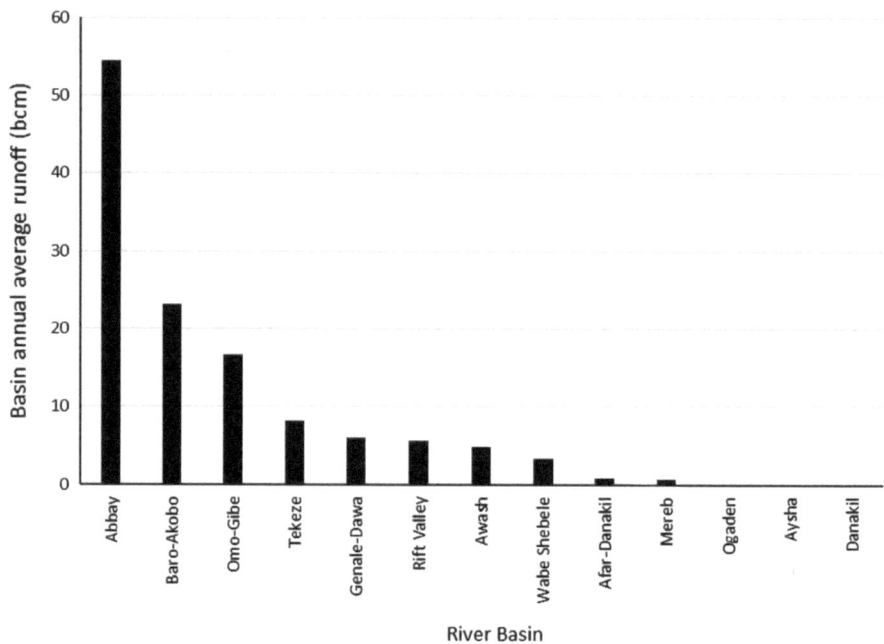

Fig. 2.3 Ethiopia's river basin and annual average flow volumes (billion cubic meter) (*Data source* Berhanu et al. 2014)

the Aswan Low Dam, a masonry gravity dam in Aswan, Egypt, that was built between 1898 and 1902, 1000 km south southeast of Cairo. The capacity of the Aswan Low Dam was 980 million m^3 (The New York Times July 27, 1913). Table 2.1 lists major dams and water control facilities on the Nile and its tributaries with hydropower, irrigation and storage capacity provided where data is available.

Large-scale dams are initiated for national economy and political benefits at the cost of long-term river dependent communities and the ecology. Looking at the number of dams on the Nile, it could look like adding few more should not start conflict. But, rather than the number of dams, it is who has full control of the dam operation is the major factor in water rights due to the direct water right implication of water control and dam operations.

2.4.1 The High Aswan Dam on The Main Nile, Egypt

The Aswan Low Dam was almost overtopped in 1946. Political changes in Egypt in the 1950s and the inadequacy of the Aswan Low Dam led to the construction of the High Aswan Dam about 4 km upstream forming Lake Nasser (Fig. 2.1). The Dam was built from 1960 to 1970. It is an embankment dam with a height of 111 m and base width of 980 m. The reservoir behind the dam, known as Lake Nasser, has surface area of 5250 km^2, maximum length of 350 and 35 km width. The reservoir capacity is 132 km^3 (132 bcm). Evaporation estimation from the High Aswan Dam reservoir based on environmental isotope distribution study was 19% of the inflow and groundwater recharge from the lake extends to 10 km distance reflecting the extent of seepage from the dam (Aly et al. 1993).

2.4.2 Sennar Dam on the Blue Nile, Sudan

Sudan has multiple water control structures and dams on the Nile and its tributaries. The sennar and Roseires are the two dams on the Blue Nile River in Sudan. The Sennar Dam is located in Sennar about 350 km south of Khartoum (Fig. 2.4). The dam has 33 m height and 3019 m length forming a reservoir 80.5 km long. It started operation on July 15, 1925 initially irrigating 122,000 ha on the Gezira plain as reported in the British Medical Journal of January 16, 1926. By 1990s, the irrigated area expanded to 860,000 ha producing mainly cotton.

2.4.3 Roseires Dam on the Blue Nile, Sudan

Roseires dam was constructed in 1966 and height and volume increased in 2013. The dam serves for power generation, flood control and irrigation. The basic current features of the dam area is a one km long concrete buttress dam of 78 m height with 24 km earthen dam on both sides, 48 m height. The reservoir behind the dam has a surface area of 627 km^2 and volume of 7.4 bcm. Over a million and half potential area is under or planned irrigation. The dam is close to the Ethiopian border and has sediment accumulation problem that led to increasing the dam height (Fig. 2.4).

2.4.4 The Grand Ethiopian Renaissance Dam on the Blue Nile River

During the construction of the High Aswan Dam in Egypt, the United States Bureau of Reclamation (USBR) studied potential hydropower sites on the Blue Nile in Ethiopia from 1956 to 1964. In 1964, it offered four proposed sites, Karadobi, Mabil, Mendia and Border (Fig. 2.5). The selection of the Border site close to the Sudan (Figs. 2.1 and 2.4) on relatively flat land requires inundation of large area to meet the required storage and water level for power generation. Details of the dam design and other aspects are presented in detail in Chaps. 6 and 7 of this book.

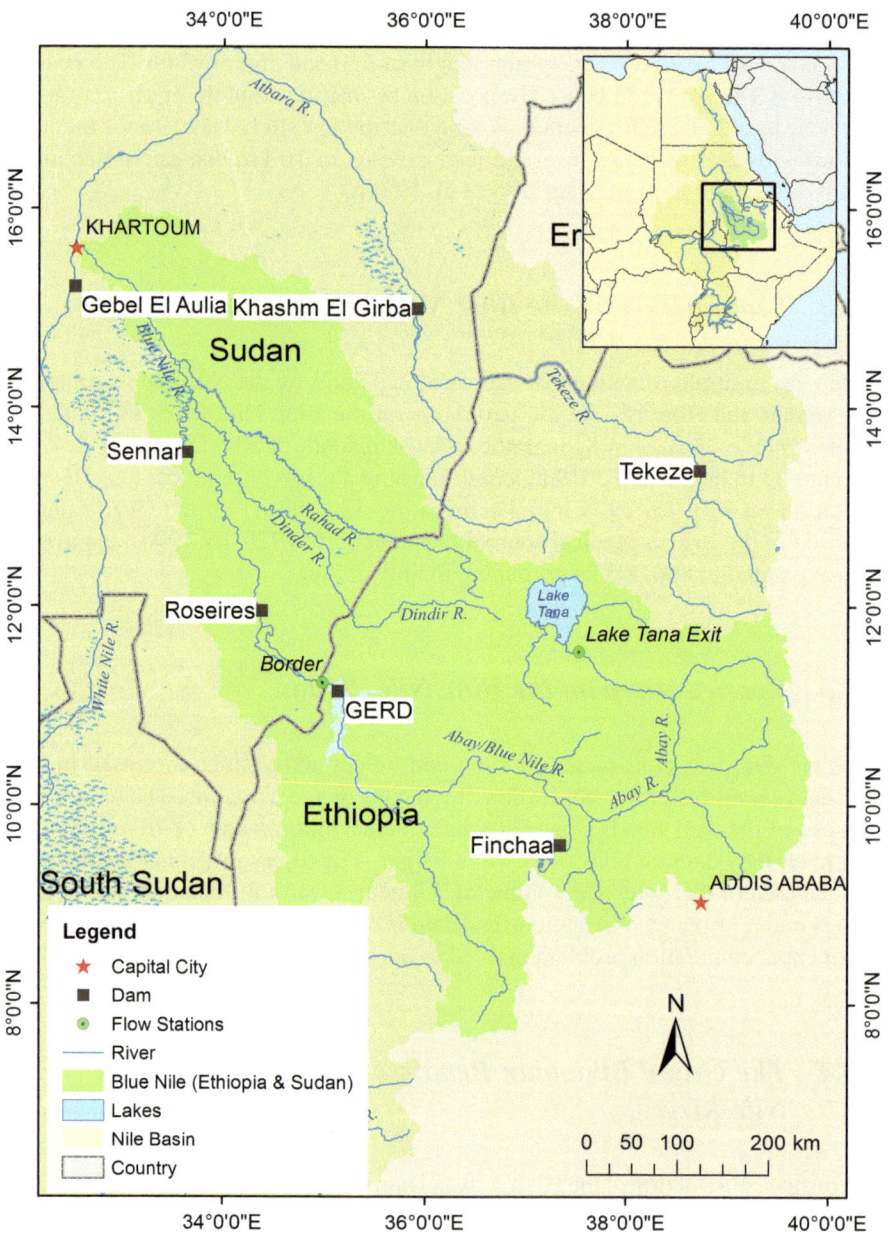

Fig. 2.4 Location of Sennar, Roseires dams in Sudan and the GERD in Ethiopia

Construction has been progressing through 2018, although initial completion was planned for 2015. By 2015, 47% of the construction was completed at a cost of 46 billion Birr ($2.16 billion). Out of this, 7.6 billion Birr ($357 million) was raised from

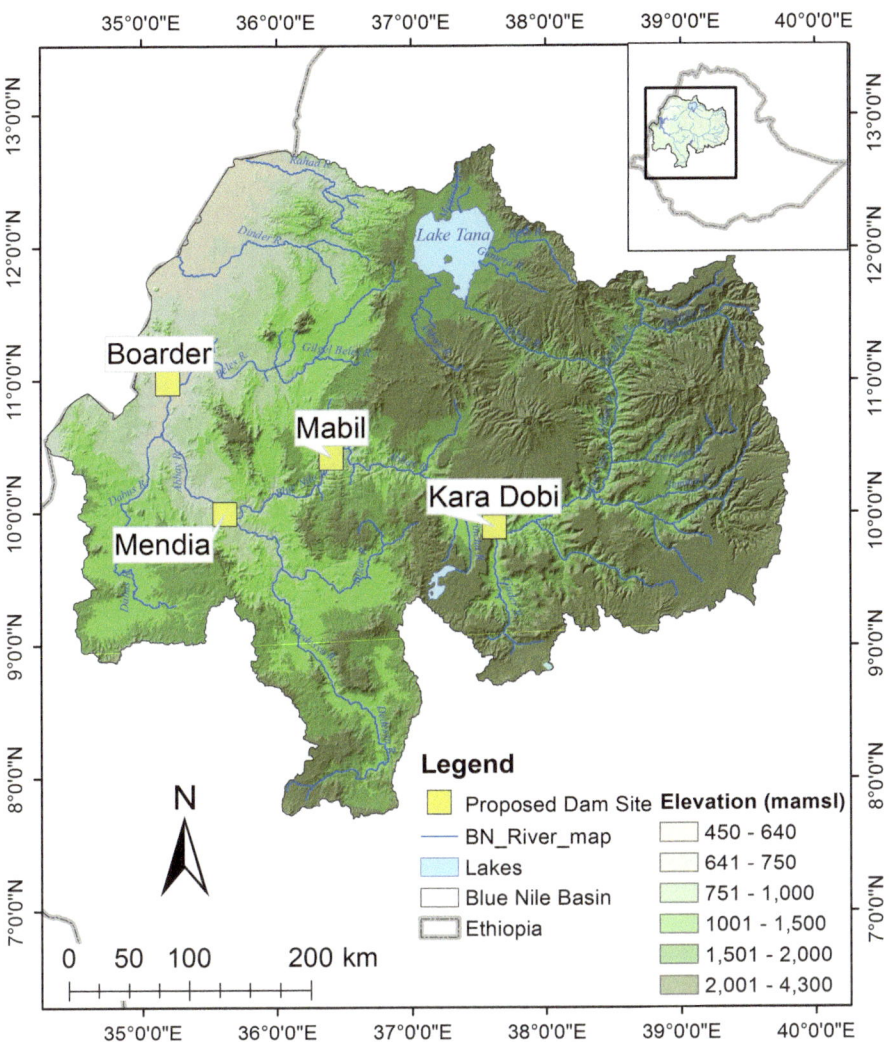

Fig. 2.5 1964 USBR proposed hydroelectric dam sites on the Blue Nile River, Ethiopia. GERD is being built near to the Boarder Site

the public while the rest is covered by the government (Ethiopian Herald 16 October 2015). At the beginning of 2018, 70% of the dam was reported to be completed by Ethiopian media. Through the construction period, several technical and diplomatic meetings underwent between Ethiopia, Egypt and Sudan to address downstream concerns. Upstream concern, the interest of people in the Nile basin that experience drought and famine at this time, has never been mentioned. Sudan in several occasions appeared mediator as it will likely benefit from the dam. Flood control, silt control, regulated water flow and power purchase will benefit Sudan. The conflicts, negotiations and agreements between the three countries are covered in Chap. 9. When the dam is complete, its success depends on the operation or regulation of water level to generate dependable hydropower. Hydrologic variation and climate change are known challenges for big dam operations from filling to normal operations in addition to downstream and upstream demands. Reduction of flows or flooding from climatic fluctuation are ripe for controversy on the construction and operation of a dam and need to be addressed before initial filling of the GERD.

2.5 Summary

Water right claims for transboundary waters are dependent on the relative military, economy, political and social strength of riparian countries. The history of water right claims in the Nile basin and link to African colonialism has been playing a major role in the present day water right claim and diplomacy. African colonial period claims of water rights are crafted by colonial powers and those arrangements continued after independence of the colonies. Post-colonial era exhibits independence and nationalism, narrowing gaps in development, population explosion coupled with water stress, and advances in water control and use. These are reflected on water right claims and water abstraction practices. Upstream and downstream water demand in the Nile River basin will continue to grow at alarming rate and water projects will keep on growing including dams and diversions, especially on tributaries. In the absence of collective agreement in water sharing and watershed management in the Nile basin, each country addresses its water needs as it perceives fit. The GERD is a reflection of the upstream countries growing need to utilize their water resources.

Climate induced reduction in flow may not be separated from upstream water use and be a cause of conflict. Water agreements, watershed management and water conservation and basin water management might be best approach to minimize water conflicts for the basin. But the pressure from water and social stresses are likely to continue as a dominant force towards unilateral water projects ultimately leading to regional conflicts. Therefore, riparian countries will need to build trust towards equitable share of the resource and its economic benefits through dialogue and understanding.

References

Abseno M (2009) How does the work of the ILC and the general assembly on the law of international water courses contribute towards a legal framework for the Nile basin. University of Dundee, Scotland, UK, Master of Laws

Abtew W, Melesse AM (2014) Ch. 28 Transboundary rives and the Nile. In: Melesse AM, Abtew W, Setegn SG (EDS) Nile River Basin ecohydrological challenges, climate and hydropolitics. Springer, New York

Aly AIM, Froehilch K, Nada A, Hamza M, Salem WM (1993) Study of environmental isotope distribution in the High Aswan Dam Lake (Egypt) for estimation of evapotranspiration of lake water and its recharge to adjacent groundwater. Environ Geochem Health 15(1):37–49

Berhanu B, Seleshi Y, Melesse AMR (2014) Surface water and groundwater resources of Ethiopia: potentials and challenges of water resources development. In: Melesse AM, Abtew W, Setegn S (eds) Nile River Basin ecohydrological challenges, climate change and hydropolitics. Springer, New York

Carlson A (2013) Who owns the Nile? Egypt, Sudan, and Ethiopia's History-changing dam. Origins 6(6)

Collins RO (1994) Ch. 5 History, hydropolitics, and the Nile: Nile control: myths or reality? In: Howell PP, Allan JA (eds) The Nile sharing a scarce resource. Cambridge University, Cambridge

Conniff K, Molden D, Pedan D, Awulachew SB (2012) Nile water and agriculture past, resent, future. In: Awulachew SB, Smakhtin V, Molden B, Pedan D (eds) The Nile River basin. Routledge, Taylor and Francis Group, New York

Degefu GT (2003) The Nile historical legal and developmental perspectives. A warning for the twenty-first century. Trafford, Victoria, Canada

El-Fadel M, El-Sayegh Y, El-Fadal K, Khorbotly D (2003) The Nile river basin: a case study in surface water conflict resolution. J Nat Resour Life Sci Edu 32:107–117

FAO (2007) http://www.fao.org/nr/water/faonile/products/Docs/Poster_Maps/BASINANDSUBB ASIN.pdf. Accessed 2 Jan 2016

Kalpakian J (2015) Ethiopia and the Blue Nile development plans and their implications downstream. ASPJ Afr Francofonie—2nd Quarter 2015

Mays LW (2010) Water technologies in ancient Egypt. In: Mays LW. Ancient water technologies. Springer, New York

McCann J (1981) Ethiopia, Britain, and negotiations for the Lake Tana Dam, 1922–1935. Int J Afr Hist 14(4):667–699

Moges SA, Gebremichel M (2014) Climate change impacts and development-based adaptation pathway to the Nile River Basin. In: Melesse AM, Abtew W, Setegn SG (eds) Nile River Basin ecohydrological challenges, climate and hydropolitics. Springer, New York

Mundy M (2015) International law approaches to Ethiopian water rights. Georgetown Environ Law Rev. (Georgetown University, Washington D.C.)

Oestigaard T (2012) Water scarcity and food security along the Nile: politics, population increase and climate change. Curr Afr (49). (Nordic African Institute, Uppsala)

Othoman AA, Rabeh SA, Fayez M, Monib M, Hegazi NA (2012) El-Salaam canal is a potential project reusing the delta drainage water for Sanai desert agriculture: microbial and chemical water quality. J Adv Res 3:99–108

Schanz M (1913) Cotton in Egypt and Anglo-Egyptian Sudan. Submitted to the 9th international cotton congress, 9–11 June, 1913. Taylor, Garnett, Evans, & Co. Manchester

Stoa R (2014) International water law principles and frameworks: perspectives from the Nile River Basin. In: Melesse AM, Abtew W, Setegn SG. Nile River Basin ecohydrological challenges, climate and hydropolitics. Springer, New York

Sutcliffe JV, Parks YP (1999) The hydrology of the Nile. IAHS, Special Publication No. 5. IAHS Press, UK

Tesemma ZK (2009) Long term hydrologic trends in the Nile basin. Masters thesis. Cornell University

Tvedt T (2004) The Nile river in the age of the British political ecology and the quest for economic power. I.B. Tauris and Co., London

Warner J (2011) Flood planning the politics of water security. New York, I.B, Tauris Co Ltd

Whole EE (2007) Hydrology and discharge. In: Gupta A (ed) Large rivers: geomorphology and management. Willey, West Sussex, England

Chapter 3
Land Tenure and Water Rights in the Blue Nile Basin

Abstract Land tenure system in the Blue Nile basin will continue to have impact on land and water resource management. The Grand Ethiopian Renaissance Dam (GERD) at the outlet of the basin will be impacted in both sediment load and basin water yield. Land is the base of political, economic, social and religious life of the Ethiopian people whether the environment is agrarian, pastoral or urban. The Ethiopian government reserves ownership of land and associated resources and administers allocation and right to use through federal and local administrative agencies. Likewise, water allocation and the right to use are not constitutionally defined but fall under other resources which are controlled by the state. Accordingly, political power and wealth in Ethiopia has been strongly tied to access and control of land. In the Blue Nile Basin, private and communal land tenure structure prevail with land ownership rights reserved for the regional and federal political powers. Lack of full ownership and responsibility of these resources has been linked to low productivity, resource degradation such as deforestation, soil and water loss. Land and water right in the basin is an important factor influencing the use of these resources by the basin dwellers. Repossessing of land through rezoning and investment justification and transferring titles make land a quick profit resource subject to growing corruption. There is no incentive for aggressive soil and water conservation watershed management.

Keywords Nile River · Blue Nile basin · Ethiopia · Water rights · Land tenure Land grab

3.1 Introduction

Land and water are the main resources for the people of Ethiopia and the Blue Nile Basin. Before the military coup led by the socialist military coalition known as DERG in 1974, the land tenure system in Ethiopia was characterized as feudal system ruled by the emperor where absentee land lords and the church had significant land holdings. Under this system, landowners collect the lion's share of the harvest from tenant

© Springer International Publishing AG, part of Springer Nature 2019   29
W. Abtew and S. B. Dessu, *The Grand Ethiopian Renaissance Dam*
on the Blue Nile, Springer Geography, https://doi.org/10.1007/978-3-319-97094-3_3

farmers. However, a large part of the Blue Nile basin had mostly small scale farmer ownership and communal land ownership tenure system which was common in the northern parts of the country. Land tenure system was the core of public uprising that overthrow the Ethiopian Monarchy and propelled the DERG to power. The DERG nationalized rural and urban land and structures. It allocated parcels agricultural lands to individual farmers effectively abolishing the feudal system. Small scale farmers were organized in collective farms and large scale farms were taken under government administration. Land transferability in form of lease, sale and inheritance was restricted. Even though farmers were able to reap their produce without sharing with landlord, there was no evidence of corresponding increase in productivity. Large scale farms were mostly underutilized as a result of inefficient government bureaucracy. After the overthrow of the military government and replacement by the Ethiopian People's Revolutionary Democratic Front (EPRDF) and associated groups. A transitional government was formed in 1991 that immediately declared continuation of land ownership of the state (Crewett et al. 2008). EPRDF re-draw the regional state map of Ethiopia based on ethnicity. The sole rural and urban land ownership provided economic power for the state ultimately helped it to project political power through absolute control of land and water resource ownership.

The current political system of Ethiopia is based on ethnic identity. The Blue Nile Basin in Ethiopia is shared by the Amhara, Oromo and Benshangul-Gumuz regions. Ethiopia launched construction of the Grand Ethiopian Renaissance Dam (GERD) on the main Blue Nile (Abay) River in the Benshangul-Gumuz region. The main source of the Blue Nile flow is the Amhara region followed by the Oromia region (Fig. 3.1), whereas the GERD reservoir will inundate more than 1700 km^2 land in the Benshangul-Gumuz region. The government has been accused by opposition groups that the ethnic based political system has played a role on the GERD site selection with respect to control of GERD and subsequent sharing of its benefits from the construction to power generation and water based economic gains of the dam (Fig. 3.2).

The impediments of state land ownership on resource conservation and initiative to maximize production have been realized in several countries. The impact of land tenure system on stunted productivity and soil loss through the three land tenure systems has been widely documented (USAID 2004). International aid donor's push for privatization of land property facing resistance from the government.

3.2 Land Rights in Ethiopia

In Ethiopia, land is the most valuable resource, natural or developed. The value of land as a source of economic and political power was realized by the current government and coded in the 1995 constitution of the country Article 40, Number 3, cementing the state's property rights (Federal Democratic Republic of Ethiopia 1995). Although, the decree states that *"land shall not be subject to sale or to other means of exchange"* but a 1996 decree that passed with a marginal parliamentary

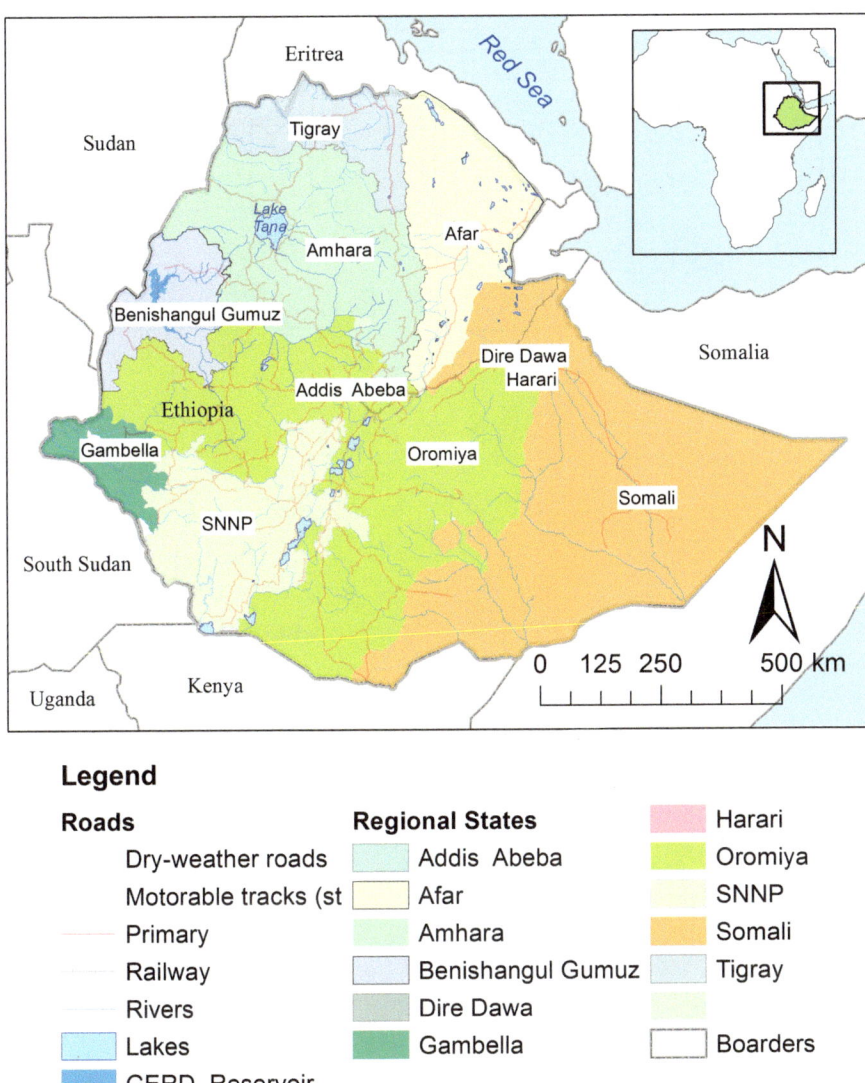

Fig. 3.1 Administrative regions/states of Ethiopia based on dominant ethnic group

vote legalized rental and leasing of land. Regional governments and branches of the Federal government were allowed to make laws pertaining to land. Land possessions are continuously transferred through urban and rural land possession reassignment for various reasons. Wide spread Land right acquisitions, displacement and land leasing for commercial farming have become source of corruption and a cause for uprisings. The Oakland institute has documented large scale land acquisitions in Ethiopia and its consequences on the people (The Oakland Institute 2011). A total of

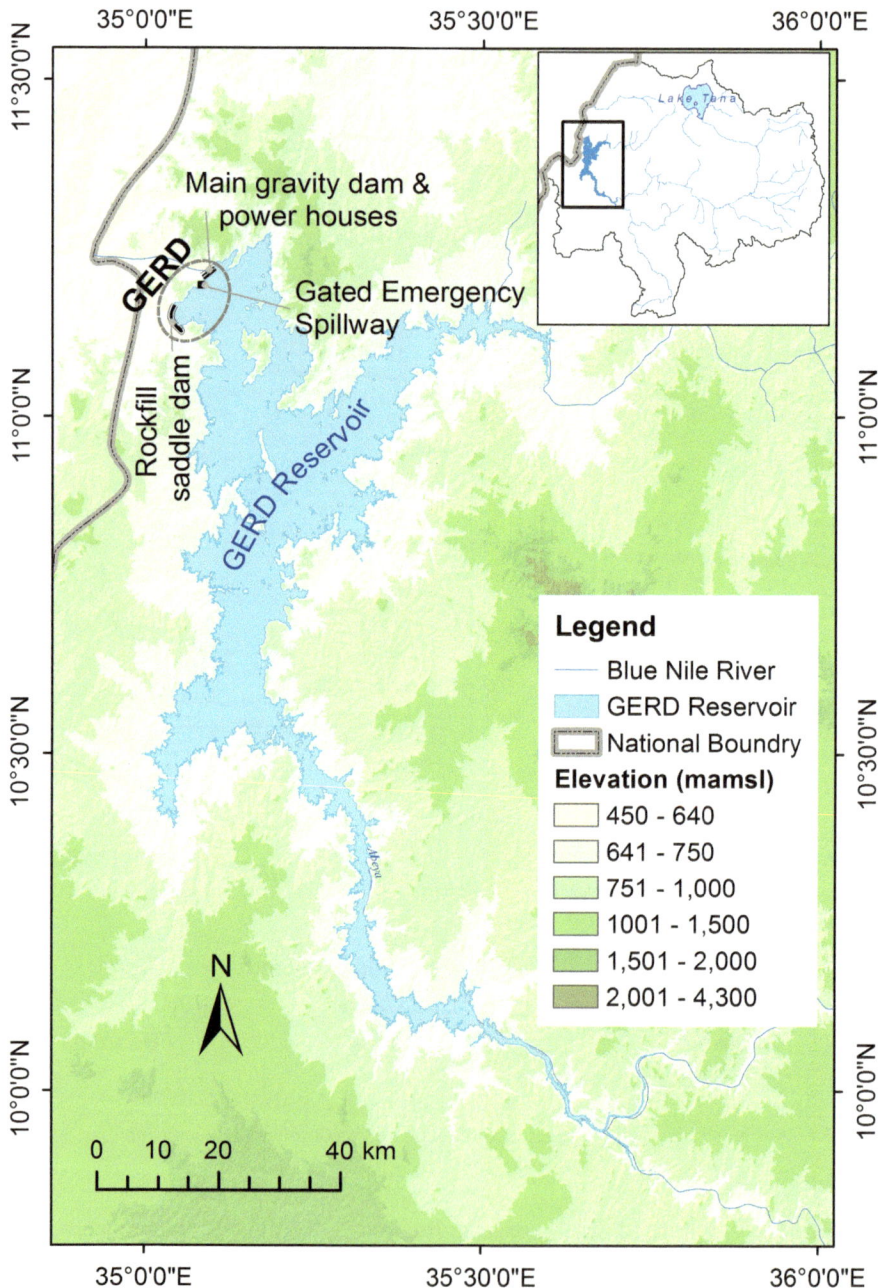

Fig. 3.2 The Grand Ethiopian Renaissance Dam at the Ethio-Sudan border

689,883 ha of land leased mostly for foreign investors has been argued as 'land grab' by local and international media. The distribution of foreign land acquisition in the Nile sub-basins was 4500 ha in Atbara (Tekeze) basin, 300,000 ha in the Blue Nile basin and 385,383 ha in Pibor-Akobo-Sobat (Baro-Akobo) basin (Liz 2016). Water demand of large scale farms require needs to be acknowledged and accounted.

Continuous issuing of new decrees pertaining to urban and rural land affirm the full right of the state on land possession and appropriations. Uncertainty in land possession rights, changing laws and corrupt judicial system has contributed to socio-political instability, loss in productivity and environmental deterioration such as in the case of flower farms. Evictions, relocations and resettlements are manifestations of the state's land rights (Pearce 2012; Lovers 2012; Legesse 2014). As land is the main resource of wealth, most corruption is associated with land. The state ownership of land has been alleged to facilitate repossession through changes in land use rezoning and other pre-eminent reasons and subsequent resale at high price; cut from land transactions; bribe from foreign investors and other forms of corruption. These allegations have been claimed by opposition groups as social injustices to urban and rural populations alike and have used them to fuel local resistance against the governing party.

Excerpt from the 1995 Ethiopian constitution that is relevant to land and water rights is given as follows, even though the constitution is not practiced in many facets.

"Article 40
The Right to Property
3. The right to ownership of rural and urban land, as well as of all natural resources, is exclusively vested in the State and in the peoples of Ethiopia.
Land is a common property of the Nations, Nationalities and Peoples of Ethiopia and shall not be subject to sale or to other means of exchange.
4. Ethiopian peasants have right to obtain land without payment and the protection against eviction from their possession. The implementation of this provision shall be specified by law.
5. Ethiopian pastoralists have the right to free land for grazing and cultivation as well as the right not to be displaced from their own lands. The implementation shall be specified by law.
6. Without prejudice to the right of Ethiopian Nations, Nationalities, and Peoples to the ownership of land, government shall ensure the right of private investors to the use of land on the basis of payment arrangements established by law. Particulars shall be determined by law.
7. Every Ethiopian shall have the full right to the immovable property he builds and to the permanent improvements he brings about on the land by his labour or capital. This right shall include the right to alienate, to bequeath, and, where the right of use expires, to remove his property, transfer his title, or claim compensation for it. Particulars shall be determined by law.
8. Without prejudice to the right to private property, the government may expropriate private property for public purposes subject to payment in advance of compensation commensurate to the value of the property".

The constitution left openings for any state action on land resulting both on urban and rural land grab through local, regional and federal level directives and decrees with cause or no cause as documented by the Oakland Institute (http://www.oakland institute.org/land-deals-africa-ethiopia); Abtew (2014); Pearce (2012) and Legesse (2014). Some of the reasons given are rezoning, urban expansion, transportation infrastructure, industrial park, "investment" and others. In urban areas the most common cause of urban land/house repossession by the government is re-zoning for high rise building. In which case, the long-time residents of a single family house were allegedly evicted from their home without fair compensation or satisfactory re-location and the land is given to developers who can secure bank loans to build high rise buildings and transfer the title. Article 89, No. 5 of the constitution solidifies the power of the state and those who are in power, on land and any other resources.

"Article 89
Economic Objectives
5. Government has the duty to hold, on behalf of the People, land and other natural resources and to deploy them for their common benefit and development."

Continuous uprising in several parts of the country has been associated with political power and economic corruption to gain land rights and by extension political rights (http://www.huffingtonpost.com/entry/ethiopia-ethnic-violence_us_56a1 0b1ee4b0404eb8f07c85).

There is evidence that continuous regional and local land rule changes create uncertainty in land possessions with defined and protected land possession rights. Some land tenure changes as land redistribution in Amhara region and reform type efforts in Tigray region are reported. Land lease and share cropping are allowed in Tigray with land ownership and land rights becoming murky (Witten 2007).

3.3 Land Rights and Resource Conservation

Major issues in land management for optimal food production and resource conservation are land to population ratio, land tenure system, social, economic, political and technological infrastructure for land right and land use management. According to Witten (2007), land rights are defined by identification of piece of land, creation of right, transformation of right, registration and preservation of right (title), and legal assurance and protection of right. Case study of the land tenure system in the Blue Nile basin and the rest of Ethiopia shows most of the population is engaged in small-scale farming. Land holdings are small compared to house hold size dependent on the land. Table 3.1 depict average land holdings in regions that are wholly and partially in the Blue Nile basin. Insecure land rights have been linked to poor resource conservation and investment on land.

Population pressure in the last 40 years has forced farmers to clear forest and steep hills to expand arable land. Lands are exposed to erosion and top soil loss of alarming rates (Asres et al. 2016; Gessesse 2014; Gashaw et al. 2014). Estimated recurring

Table 3.1 National and regional percentage of land holding per farmer in the regional states of the Blue Nile basin (Nega et al. 2003)

Farm size (ha)	National (Ethiopia)	Amhara	Oromia	Benshangul
Landless	10	9.8	13.6	14.4
<0.5	27.6	40.3	17.8	–
0.51–0.75	13.1	19.1	11.5	3.6
0.76–1.0	12	9.4	11.9	13.5
1.01–1.50	14	14.2	15.1	13.5
1.51–2.00	8.1	3.5	11	10
2.01–3.00	11.5	3.3	13.9	26.1
3 ha or more	3.7	0.4	5.2	18.9
Average holding (ha)	1.02	0.75	1.15	1.82
Average land-labor ratio	0.38	0.3	0.4	0.64
Random sample of farmers	8540	1703	3905	122

cost of soil loss in Ethiopia is 2–3% of annual GDP (World Bank 2006). Yesuf et al. (2005) reported a 3% annual GDP loss due soil erosion with severe case in eastern Blue Nile (Wollo region) and Northern Blue Nile (Gonder). A USAID study (2004) summarizes impact of land tenure system in Ethiopia covering the Blue Bile basin as follows *"Research and studies in Ethiopia show that insecurity of land tenure restricts rights in land, reduces incentives to productively invest in land, and limits transferability of land. In return, these pose significant constraints to agricultural growth and natural resource management."*

3.4   Water Rights and Water Conservation

Water, as a natural resource, is presumably under the state's jurisdiction according to Article 89 of the Ethiopian constitution. Small scale water withdrawal from streams and rainfall harvesting are unregulated except local social structures are formed to ensure local accord among competitive users. A case study of small scale irrigation local water resources governance structures in Fogera and Guba Lafto Woredas around the perimeter of Lake Tana in Amhara Region are reported (Deneke 2014). Water Use Associations (WUA) are established based on regional cooperative establishment proclamation and irrigation cooperatives establishment directive. The WUAs administer water allocation and resolve water conflicts among users and stakeholders. While the associations manage gravity flow diversion structures, financially capable farmers introduced pumps and operate outside the association's oversight creating water shortage in some cases.

(a)   (b)

Fig. 3.3 a Water harvesting pool digging in the Blue Nile basin upstream of GERD; **b** Water harvesting pool with plastic liner in the Nile basin

Food shortage and lack of drinking water is common in many areas in the Blue Nile basin. The eastern Amhara region is drought prone and water scarce where communal water harvesting ponds are necessity. Shallow wells, springs and streams are source of water supply (Deneke 2014). There was a national program of local water harvesting through catching runoff in ponds to relief food and water shortage in the Nile basin in Ethiopia that was aggressively expanded (Fig. 3.3a, b). However, how much the program is currently supported is not clear. Benefits and associated problems of water harvesting mainly in the form of ponds has been reported (Tesfay 2011; Mume 2014).

The Blue Nile and other Ethiopian rivers are trans-ethnic within Ethiopia with potential internal water conflict as much as with downstream users outside Ethiopian border. The dam has a potential to be a cause of conflict within Ethiopia for land, water, power and reservoir associated developments and upstream water rights as long as the current ethnic based constitution is enacted. Upstream water rights are in conflict with the dam objectives and no recognition of the issue and consultation of upstream people is reported. Growing upstream dependence on irrigation and water abstraction could soon surface as water right issue.

For large scale farms, it appears the possession of land close to a water resource, by default grant the right to use the water with no obligation of water conservation. This large farms can be state sugar farms in Omo valley, Awash or Belles; individual or business run national or foreign flower farms and other farms in various parts of the country including the southwest fertile region.

3.5 Summary

Land tenure system has direct bearing on productivity, economy, optimal use of resources, and soil and water conservation. State ownership of land and water resources has cemented power for the state while creating societal instability. Land reform securing land rights may promote environmental sustainability and soil conservation reducing the current alarming soil loss in the Blue Nile basin upstream of GERD. Growing small and large scale irrigation developments need to be factored in water resource planning and management. Sound water management policy and guideline are essential for long-term water conservation and optimal returns from land and water resources of the Blue Nile basin.

References

Abtew W (2014) Chapter 7: Land and water in the Nile basin. In: Melesse AM, Abtew W, Setegn SG (eds) Nile River Basin ecohydrological challenges, climate change and hydropolitics. Springer, New York

Asres RS, Tilahun SA, Ayele GT (2016) Chapter 5: Analysis of land use/land cover change dynamics in the upland watersheds of the Upper Blue Nile Basin. In: Melesse AM, Abtew W (eds) Landscape dynamics, soils and hydrological processes in varied climates. Springer, New York

Crewett W, Bogale A, Korf B (2008) Land tenure in Ethiopia: Continuity and change, shifting rulers, and the quest of state control. CAPRi Working Paper No. 91. CAPRi

Deneke TT (2014) Chapter 24 Processes of institutional change and factors influencing collective action in local water resources governance in the Blue Nile Basin of Ethiopia. In: Melesse AM, Abtew W, Setegn SG (eds) Nile River Basin ecohydrological challenges, climate and hydropolitics. Springer, New York

Federal Democratic Republic of Ethiopia (1995) The constitution of the Federal Democratic Republic of Ethiopia. Addis Ababa, Ethiopia

Gashaw T, Bantider A, G/Selassie H (2014) Land degradation in Ethiopia: causes, impacts and rehabilitation techniques. J Environ Earth Sci 4(9):98–104

Gessesse GD (2014) Chapter 11: Assessment of soil erosion in the Blue Nile Basin. In: Melesse AM, Abtew W, Setegn SG (eds) Nile River Basin ecohydrological challenges, climate change and hydropolitics. Springer, New York

Legesse E (2014) Yemelese trufatoch balebet alba ketema (Amharic). Netsanet Publishing Agency

Liz EC (2016) Water grabbing and conflict in the Nile River basin: a focus on Ethiopia. MS thesis. University of British Columbia, Vancouver

Lovers T (2012) 'Land grab' as development strategy? The political economy of agricultural investment in Ethiopia. J Peasant Stud 39(1)

Mume J (2014) Impact of rain-water-harvesting and socio economic factors on household food security and income in moisture stress areas of eastern Hararghe, Ethiopia. Int J Novel Res Market Manage Econ 1(1):10–23

Nega B, Adenew B, Gebre Sellasie S (2003) Current land policy issues in Ethiopia. Land Reform, Land Settl, Cooperatives 11(3):103–124

Pearce F (2012) The land grabbers: the new fight over who owns the earth. Beacon Press, Boston, MA investment

Tesfay G (2011) On-farm water harvesting for rainfed agriculture development and food security in Tigray, Northern Ethiopia. DCG Report N0:61

The Oakland Institute (2011) Understanding land investment deals in Africa country: Ethiopia. Oakland, Ca

USAID/Ethiopia (2004) Ethiopia land policy and administration assessment. Final report with Appendices, Submitted by ARD, Inc, Burlington, VT

Witten MW (2007) The protection of land rights in Ethiopia. Africa Focus 20(1–2):153–184

World Bank (2006) The cost of land degradation in Ethiopia: a review of past studies. World Bank Paper 61128

Yesuf M, Mekonnen A, Kassie M, Ponder J (2005) Cost of land degradation in Ethiopia: a critical review of past studies. Environ Econ Policy Ethiopia, December 2005

Chapter 4
Hydrology of the Blue Nile Basin Overview

Abstract The Blue Nile (Abay) basin contributes close to 60% of the flow of Nile River while draining only 10% of the Nile Basin. Analysis and study of the past, current and future climate and hydrology of the Blue Nile basin is important to sustain the environment and livelihood in the Nile Basin. The basin mean annual rainfall is 1423 mm with 74% in the four wet months of June to September. Rainfall in the basin is influenced by ENSO events where El Niño years are likely to be drier than normal and La Niña years are likely to be wetter. River flow fluctuates with the timing and amount of rainfall. The estimated mean annual flow of the Blue Nile River at the Ethio-Sudan border is about 50 bcm (billion cubic meter) with range of year-to-year variation. The subsistence rainfed agriculture in the basin has been affected by periodic droughts and may not be sustainable to support the growing population. Increasing water demand in the Nile basin for food and energy is manifested in the continual increase of water control projects on the main rivers and tributaries and expansion of irrigated areas. These projects are likely to alter the hydrology of the Blue Nile Basin. One of the currently undergoing projects is the Grand Ethiopian Renaissance Dam (GERD) being built on the Blue Nile River by Ethiopia for hydroelectric power generation. This chapter provides an overview of the hydrology of the Blue Nile Basin.

Keywords Blue Nile river basin · Blue Nile river basin hydrology
Dams in the Blue Nile river basin · Grand Ethiopian Renaissance Dam

4.1 Introduction

Ethiopia has twelve drainage basins with three dry basin; Aysha, Danakil and Ogaden (Table 4.1, Fig. 4.1). The nine basins generate an estimated annual surface water flow of 124.4 bcm of which 44% is from the Blue Nile basin (Berhanu et al. 2014). With the exception of Awash River, every river in Ethiopia is transboundary. The Awash drains to Lake Abbe in the Afar depression at the Ethio-Djibouti border (Dessu et al. 2016). The western basins flow to the Sudan, the southern basins to Kenya

© Springer International Publishing AG, part of Springer Nature 2019
W. Abtew and S. B. Dessu, *The Grand Ethiopian Renaissance Dam on the Blue Nile*, Springer Geography, https://doi.org/10.1007/978-3-319-97094-3_4

Table 4.1 Ethiopia's major river basins, watershed area and flows (MoWR 2016)

No.	Basins	Catchments area		Annual discharge	
	%	(km^2)	%	Billion m^3	%
1	Abay[a]	199,812	17.56	54.4	43.05
2	Awash	112,700	9.9	4.9	3.76
3	Baro-Akobo	74,102	6.51	23.23	19.31
4	Genele-Dawa	171,050	15.03	6.1	4.81
5	Tekeze	90,000	7.9	8.2	6.24
6	Wabi-shebele	200,214	17.59	3.16	2.59
7	Omo-Ghibe	78,200	6.87	16.6	14.7
8	Mereb	5900	0.52	0.72	0.21
9	Rift valley Lakes	52,740	4.63	5.64	4.62
10	Danakil	740,002	6.5	0.86	0.7
11	Ogaden	77,100	6.77	0	0
12	Aisha	2200	0.19	0	0
	Total	1138,020	100	123.81	100

[a]Blue Nile

and the eastern basin to Somalia. In response to periodic drought and water shortage associated with climate and hydrology, Ethiopia has been increasing utilization of its water sources to satisfy the growing food and energy demand.

The Blue Nile basin hydrology is very important to millions who currently and in the future depend on its water for their survival. The Blue Nile basin covers 199,812 km^2 surface area. The drainage area for contributing flow to GERD reservoir is 172,250 km^2. The annual flow of the Blue Nile is reported as 54.4 bcm, with 746 m^3 d^{-1} km^{-2} runoff generation rate (Berhanu et al. 2014). Eighty six percent of the annual flow of the Nile comes from Ethiopia with contributions of the Upper Blue Nile Basin (59%), the Baro–Akobo–Sobat sub-system (14%) and 13% from the Tekeze/Atbara/Gash sub-system (Degefu 2003). Annual rainfall ranges from less than a 900 mm in the Northeast to over 2000 mm in the south. On the basis of annual rainfall and river discharge, the Blue Nile basin is the wettest part of Ethiopia.

The tributaries of the Blue Nile River are Woleqa, Beshlo, Jemma, Muger, Guder, Chemoga, Finchaa, Didessa, Beles, Angar, Dabus, Gilgel Abay, Ribb, Gumera, Wenchit and numerous streams. Runoff is generated from urban, agricultural, forest and other undevelopable areas. Streams which flow through settlements are currently and potentially useful for small scale and medium scale irrigation through dams and diversion structures. Runoff from escarpments and gorges that flow into major tributaries or the Blue Nile itself require major dams on the tributaries and the Blue Nile to access the water for major irrigation projects and hydropower.

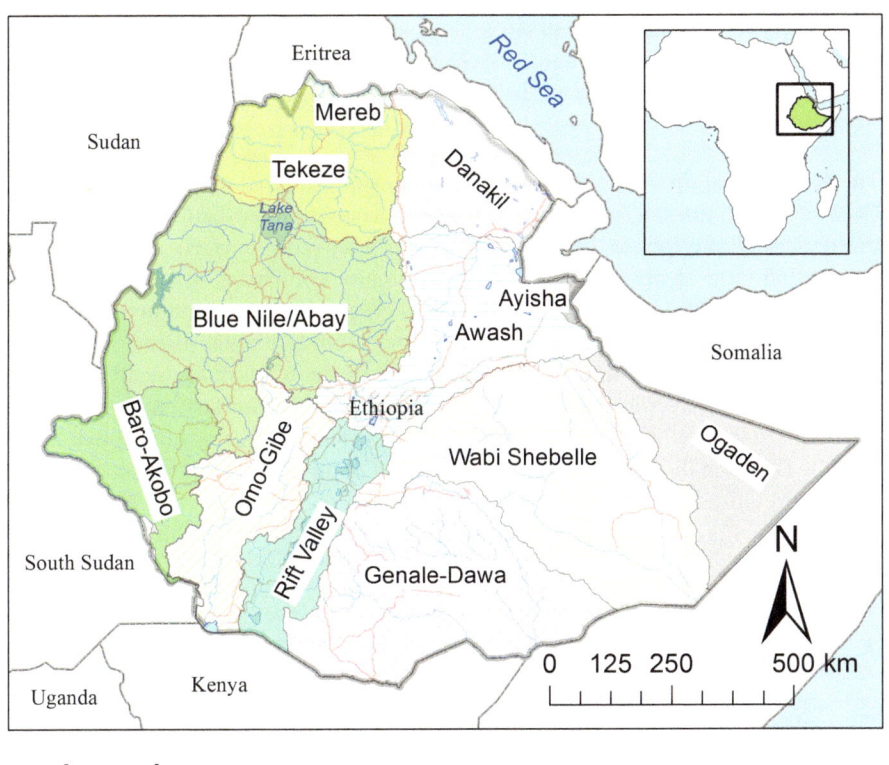

Fig. 4.1 The twelve surface water basins of Ethiopia

The four western basins, Blue Nile/Abay, Mereb, Tekeze, and Baro-Akobo are part of the Nile basin flowing to Sudan and South Sudan. The Blue Nile basin is the largest basin contributing about 48–50 bcm of water at the Ethio-Sudan boarder. The three Eastern basins are dry basins.

4.2 Physical Characteristics

4.2.1 Topography

The Gelgel Abay in the right bank of the main stem starts at the Choke Mountain in Gojam with an altitude of above 3800 m asl. The Gummera and Rib rivers (major tributaries of Lake Tana in the east) start flowing from the Guna peak, in south Gondar, with an altitude of above 4000 m asl. The major tributaries in its left bank start flowing from the central highland plateaus of north Showa and southern Wollo with an average altitude of above 2200 m asl. In the south left bank, its major tributaries (Guder, Finchaa, Angar and Deddessa) flow from the western highland plateaus of Ethiopia with an average altitude of above 2000 m asl. Similarly tributaries in the right bank start flowing from the Gojam highlands with peaks reaching above 3000 m asl. The Dabus in the left bank and the Rahad & Dindir in the right bank starts flowing in the mid altitude plateaus (about 2000 m asl) of west Wellega and North Gondar respectively. Figure 4.2 indicates altitude variations in the Blue Nile sub basin.

4.2.2 Land Slope

The total area of the Blue Nile basin is estimated at 199,812 km^2. The watershed with less than 5% land slope covers 44%, and land slope ranging from 5 to 10% covers 10% of the sub basin. Land slope in the range of 10–15% and 15–20% cover 12 and 10% of the sub basin respectively. High land slope areas (>20%) with rugged and undulating land surfaces of non-uniform topography covers 24% of the sub basin. Mild slope area of the sub basin (<5%) is largely located in the lower course of the sub basin, with some area located at the foot of high mountains and highland plateaus (1800–2400 m asl) in its upper course (Fig. 4.3).

4.2.3 Soils

The Blue Nile basin in Ethiopia is dominantly covered with clay soil. Vertisols are estimated to cover 15% of the basin mostly in the poorly drainage areas. On better drained sites of moderate slope, especially in the areas of high rainfall or in the areas of long stability (the lowlands), red soils dominate (Alisols, Acrisols, Nicosols) covering some 42% of the basin. The other main productive soils are Luvisols occupying 9% of the basin. About 32% of the basin is covered by shallow and moderately deep soils, including Leptospls (21.5%), Cambisols (9.5%), Regosols and Artenosols (1.3%). Figure 4.4 depicts the soils of the Blue Nile.

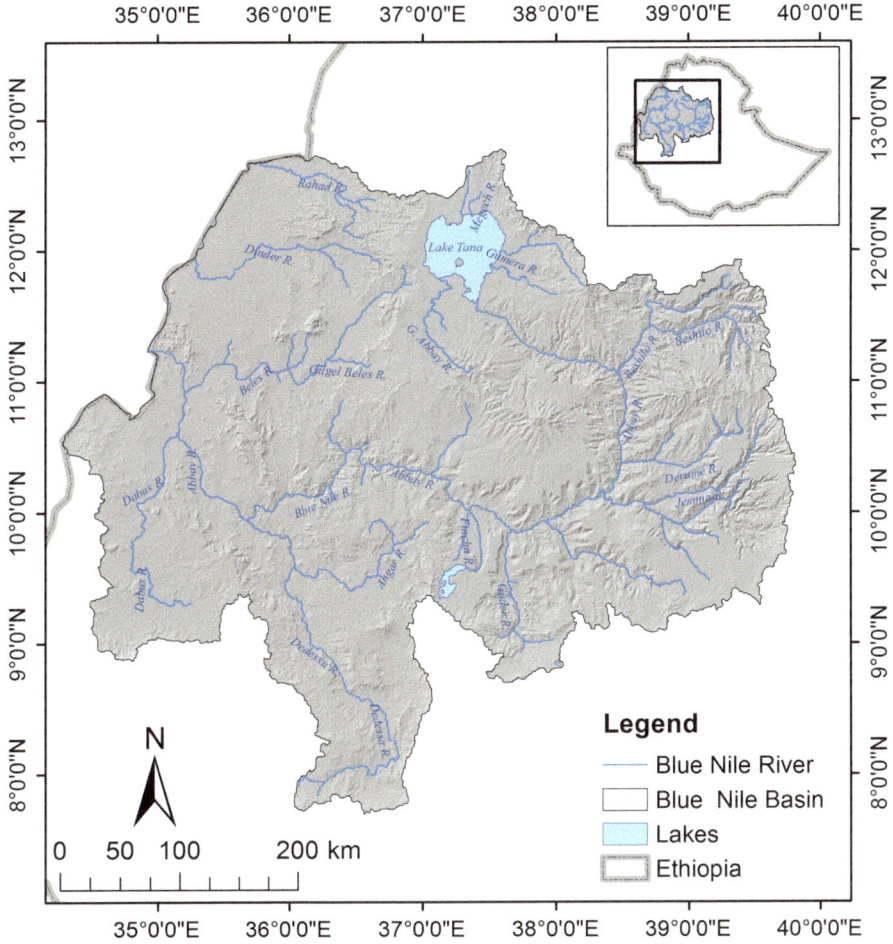

Fig. 4.2 Topography and major tributaries of the Blue Nile derived from a 30 m resolution DEM from the Shuttle Radar Topographic Mission (Jarvis et al. 2008)

The slope and soil type contribute to the high sediment load on the Blue Nile River. The upper Blue Nile River, especially the Lake Tana Basin, has relatively flat terrain and productive soil suitable for irrigation.

4.2.4 Land Cover

The land cover of the basin essentially follows the divide between highland and low land. The highlands were converted from once dominantly sub-humid tropical forests to cultivation and grazing. Almost the entire highland area is under farmland

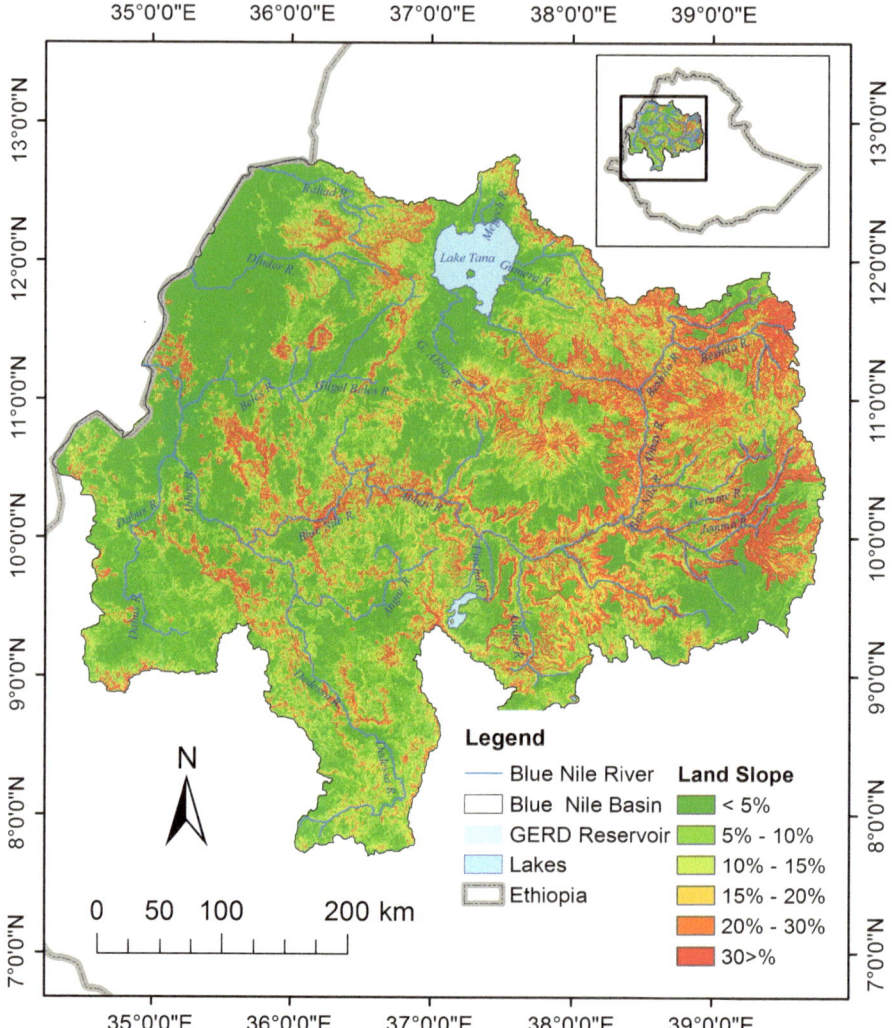

Fig. 4.3 Land slope in the Blue Nile basin

(33.9%). Forest remnants remain in the south-west. Other forest remnants exist along the highland lowland divide, generally on the steep slopes. The Forest cover is about 1.41% of the basin area. The other major highland land cover is grass land (23.1%), which occurs primarily either in poorly drained depressions or on level (poorly drained) and exposed high altitude locations. Wood lands and Bush and Shrubs cover 20.3 and 10.2% of the Abay Basin area. Figure 4.5 depicts land cover of the Blue Nile basin.

The Blue Nile basin land use/cover is mostly Agriculture with scanty forest areas. The traditional rainfed agriculture contributes to sediment load in the basin. However,

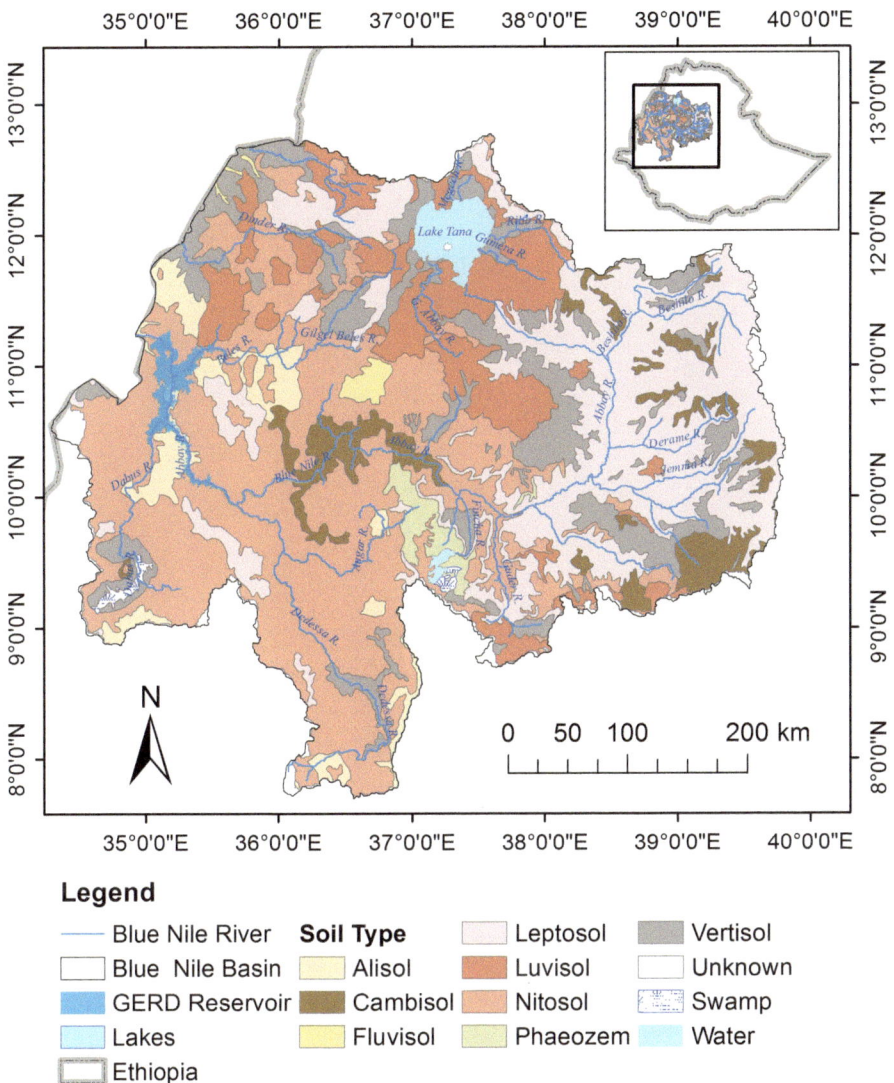

Fig. 4.4 FAO, 1:5 million harmonized world soil database (Fischer et al. 2008; FAO et al. 2009)

there is an increasing trend of small scale irrigations mostly on the upstream reaches of the Blue Nile Basin. The GERD will increase the coverage of water bodies and reduces the Agricultural land cover.

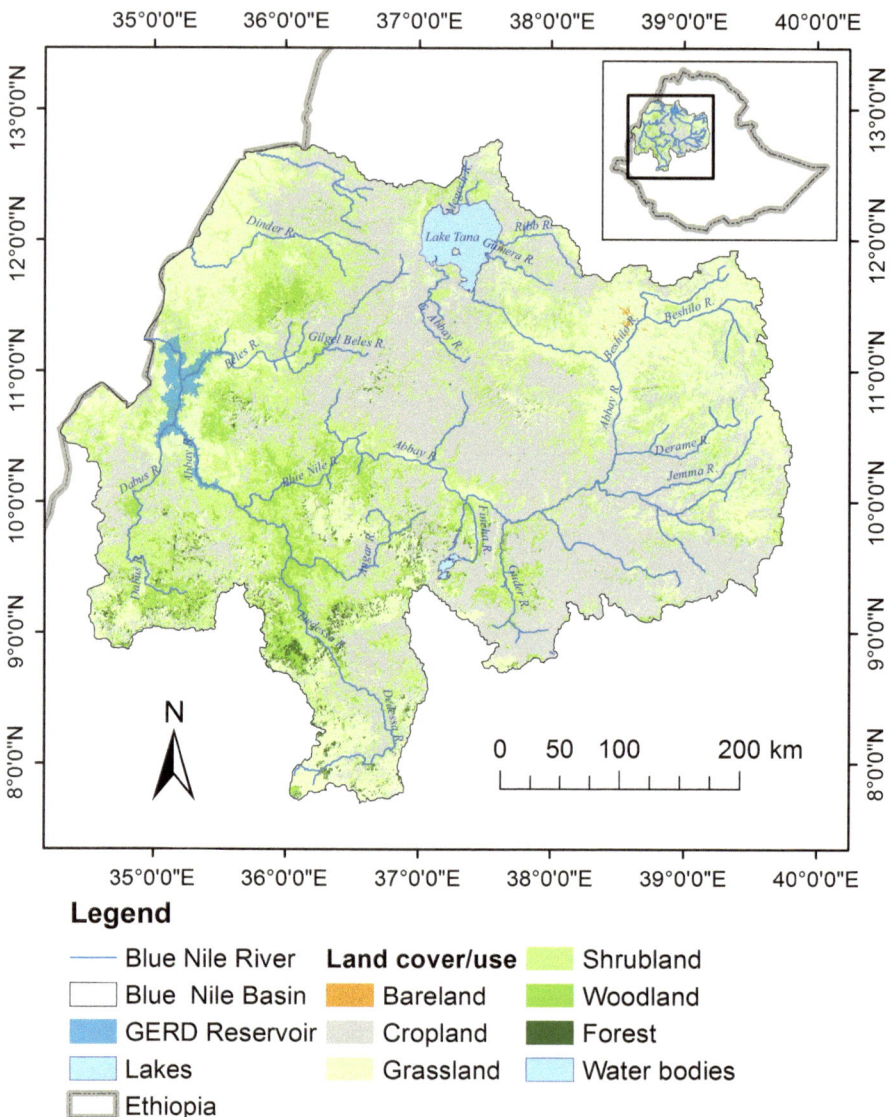

Fig. 4.5 300-m resolution land use/cover raster data (ESA 2008) of Blue Nile basin

4.3 Climate

4.3.1 Blue Nile Basin Climate

The Climate of Ethiopia and the Blue Nile basin is mostly regulated by altitude as the country is in tropical latitude. High elevation areas (>3200 m) are classified Wurch, cold; cool areas (2300–3200 m) are classified Dega; mild areas (1500–2300 m) are classified as Woyna dega; warm areas (500–1500 m) are classified as Kola and hot areas (<500 m) are classified as Berha (Berhanu et al. 2014). All five climatic zones exit in the Blue Nile basin with mostly Dega and Woyna dega on the plateau and kola along the Sudan border (Fig. 4.6).

The eastern highlands are mostly Wurch whereas the western lowlands are Kola. The GERD reservoir will be completely in the Kola climate. The Woyna Dega climate zone is relatively populated area of the basin.

Meteorological data summary for three sites in the Blue Nile basin from north to south are depicted in Table 4.2a–c. Monthly average temperature (T), relative humidity (H), wind speed (W/S) and sunshine hours (h) are shown. In the absence of solar radiation data, average potential evapotranspiration (ETo) is estimated from average temperature and sunshine hours using the Blaney-Criddle method (Abtew and Melesse 2013). Bahir Dar (11.6°, 37.38°) is at elevation of 1801 m. Debre Markos (10.33°, 37.72°) is at an elevation of 2447 m. Asosa (10.14°, 34.74°) is at an elevation of 1570 m. Estimated annual ETo are Bahir Dar (1664 mm); Debre Markos (1521 mm) and Asosa (1772 mm).

4.3.2 Rainfall in the Blue Nile Basin

The Nile Basin rainfall is driven by land-ocean-atmosphere interactions and the Inter Tropical Convergence Zone. The ITCZ is the zone of convergence of winds from the southern and northern hemispheres. It swings north-south along the equator affecting the timing and pattern of rainfall in the Nile basin (Glantz 1987; Dessu and Melesse 2013). The migration of ITCZ is sensitive to variations in Indian Ocean sea surface temperatures that vary from year to year influencing the onset, duration and intensity of rainfalls in the Blue Nile Basin as well as episodes of El Nino and La Nina (Abtew et al. 2009b). Rainfall in the Blue Nile Basin varies spatially and temporally. Rainfall in the Blue Nile basin varies spatially from 900 mm at northeast to 2000 mm at the south boundary of the basin (Fig. 4.7). Seasonally, November through April is the dry season with less than 60 mm average rainfall. December, January and February have less than 15 mm per month of rainfall. October is the transition month into the dry season while May is the transition into the wet season (June through September).

Spatial characteristics of rainfall in the Blue Nile basin is important from the perspective of rainfed agriculture, irrigation and hydropower. Rainfed agriculture is

Table 4.2 Monthly meteorological variables

Month	Temperature	Humidity	Wind speed	Sunshine	ETo
	°C	%	m s^{-1}	h	mm d^{-1}
(a) At Bahir Dar					
Jan	16.6	57	1.7	9.7	4.22
Feb	18	50	2	9.5	4.4
Mar	20.4	47	2.1	9	4.69
Apr	20.9	46	2	9.2	4.93
May	21.3	57	2.2	8.2	4.98
Jun	20	65	2.5	7.1	4.82
Jul	18.6	78	2.1	5.3	4.64
Aug	18.4	80	1.7	4.8	4.45
Sep	18.7	74	1.5	6.6	4.48
Oct	18.9	68	1.1	8.5	4.51
Nov	17.9	63	1.1	9.5	4.38
Dec	16.3	59	1.3	9.5	4.18
(b) At Debre Markos					
Jan	15.8	48	1.2	9.6	4.12
Feb	16.8	47	1.2	9	4.25
Mar	17.7	50	1.3	8.3	4.36
Apr	17.8	54	1.2	7.3	4.53
May	17.2	62	1.2	7.8	4.46
Jun	15.2	79	1.1	6.1	4.2
Jul	14.3	88	0.9	3.6	4.08
Aug	14.3	87	1	3.7	3.94
Sep	14.6	81	1	6.4	3.97
Oct	15	63	1.3	8.6	4.02
Nov	15	57	1.1	8.5	4.02
Dec	15.1	52	1.2	9.8	4.04
(c) At Asosa					
Jan	21.8	58	1.7	8.4	4.87
Feb	23.2	53	2	8.8	5.04
Mar	23.9	58	1.2	8.3	5.13

(continued)

Table 4.2 (continued)

Month	Temperature	Humidity	Wind speed	Sunshine	ETo
	°C	%	m s^{-1}	h	mm d^{-1}
Apr	23.3	64	1.5	7.7	5.24
May	22	74	1.5	5.6	5.07
Jun	20	82	1.4	5.2	4.82
Jul	19.2	87	1.3	3.8	4.71
Aug	19.2	88	1.1	3.7	4.54
Sep	19.7	86	1.3	4.5	4.61
Oct	20	85	1.4	5.5	4.64
Nov	20.8	75	1.4	8.5	4.74
Dec	21.5	59	1.7	8.8	4.83

subject to rainfall fluctuation in the basin while stream flow fluctuation is a major concern for hydropower and irrigation.

Seasonal variation of rainfall pattern at a location or on a region and its cyclic impact on soil moisture and stream flow is the foundation of livelihood in the Nile basin. Rainfed agriculture is totally dependent on the dry and wet cycling patterns for planting and harvesting of crop. The cyclic seasonal flooding and recession of the Nile has created adaptations of survival downstream. The wet season in the Blue Nile Basin is June, July, August and September. The dry season is November through April. May is the transition month from dry to wet while October is the transition from wet to dry season (Fig. 4.8).

Temporal variation of rainfall is the variation of rainfall from year to year for each month and each season. This variation determines droughts and floods or close to average conditions. Monthly rainfall distribution fits a number of theoretical probability distributions (Gamma, Normal, Log-Normal and Weibull) as reported in Abtew et al. (2009b). Rainfall variation is high in the dry season followed by the transition months. Wet season rainfall variation is lower indicating that chances of getting no rainfall being very small during this season (Fig. 4.8).

Blue Nile basin average rainfall temporally varies between 1100 and 1700 mm based on 32 gauge network from 1960 to 2002, Fig. 4.9. The mean annual rainfall is 1423 mm with standard deviation of 125 mm.

Monthly rainfall return periods show the probabilities of getting a certain amount of monthly rainfall based on frequency distributions. As commonly presented in the literature, rainfall for various return period from 100-year dry to 100-year wet are estimated. Figure 4.10 depicts monthly rainfall estimates for various return periods for the dry season months (November, December, January, February, March, April); transition months (May and October) and wet season months (June, July, August, September). The transition months of May and October have ranges of rainfall probabilities between the dry and wet season months (Fig. 4.10).

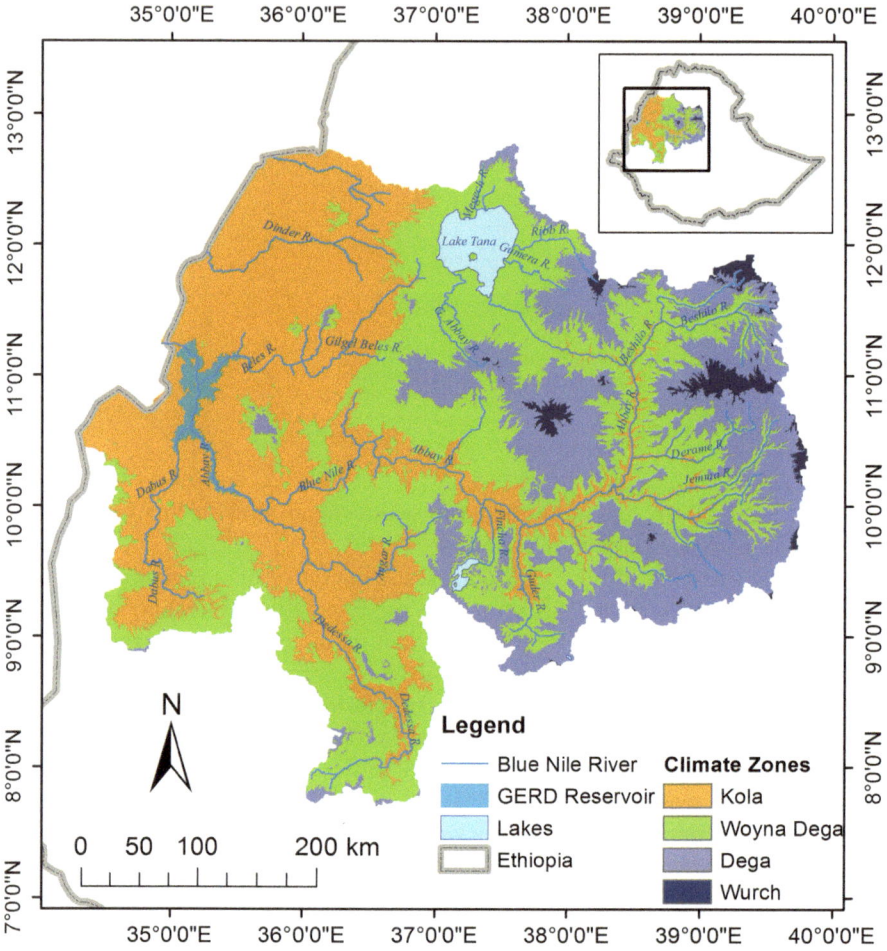

Fig. 4.6 Climatic zones of the Blue Nile basin based on Elevation

4.3.3 *Effect of ENSO Events on the Blue Nile Rainfall*

Study of yearly Nile low flow and flood levels data observed by ancient Egyptians, using the Roda Nilometer at Cairo, Egypt, showed flow variation between 622 and 1470 AD (Putter et al. 1998). Spectral peaks of various years were reported and the 5-year period in the flood stage over most of the study period was postulated as a robust link between ENSO (El Niño Southern Oscillation) and Nile discharge. From tree ring history and recent instrumental observations (500–2000 AD), the occurrence of El Niño at 2–7 years interval was shown along with El Niño amplitude identifying strong El Niño events at different periods (IPRC 2011). The strong El Niño peaks at 800 AD, 1250 AD, 1400 AD match with the Nile low water level record. Parallel

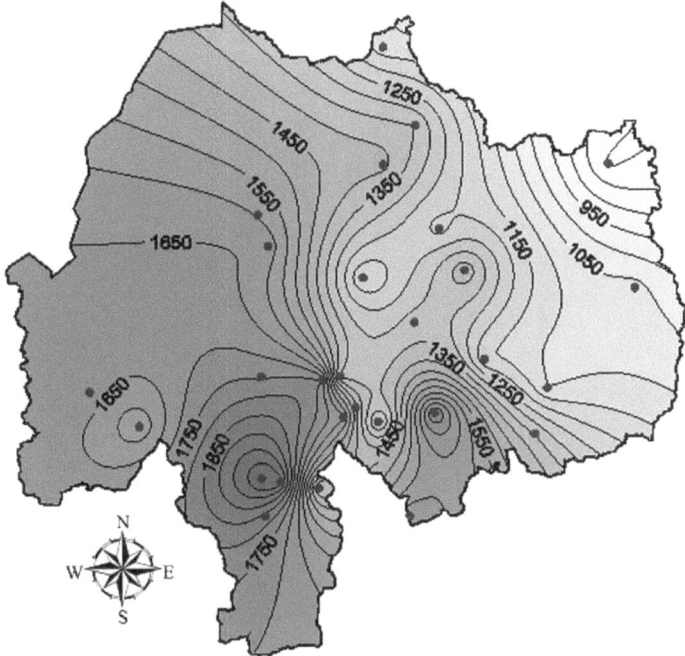

Fig. 4.7 Spatial variation of average annual rainfall on the Blue Nile basin (Abtew et al. 2009b)

conclusions were derived from sediment core studies from Lake Turkana (Halfman et al. 1994).

Recent period studies have also demonstrated the link between Ethiopian rainfall and ENSO. Seleshi and Zanke (2004) reported that June–September rainfall of the Ethiopian highlands is positively correlated to the Southern Oscillation Index (SOI) and negatively correlated to the equatorial eastern pacific SST. Positive SOI indicates La Niña condition while negative SOI indicate El Niño condition. Gissila et al. (2004) developed linear regression equations to forecast Ethiopian summer rains with variables that included sea surface temperature (SST) anomalies of the western Indian Ocean, the tropical eastern Indian Ocean and Niño 3.4 SST anomalies for the preceding March, April and May rainfall. Semazzi et al. (1996) study of teleconnections between global SST anomalies and rainfall over the Sahel and Southern Africa, with a Global Circulation Model (GCM), showed that externally forced SST variability dominates over internal variability in explaining the 1973 drier conditions. Abtew et al. (2009a) showed that Blue Nile basin rainfall is related to ENSO with above average likely during La Niña years and drought during El Niño years. Major Ethiopian droughts correspond to strong El Niño Years. The 1888–1892 great famine (Pankhurst 1966) correspond to the strong El Niño of 1888. Seleshi and Zanke (2004) reported that 1965, 1972–1973, 1983–1984, 1987–1988 and 1997 were drought years with low agricultural production and societal impact. These were El

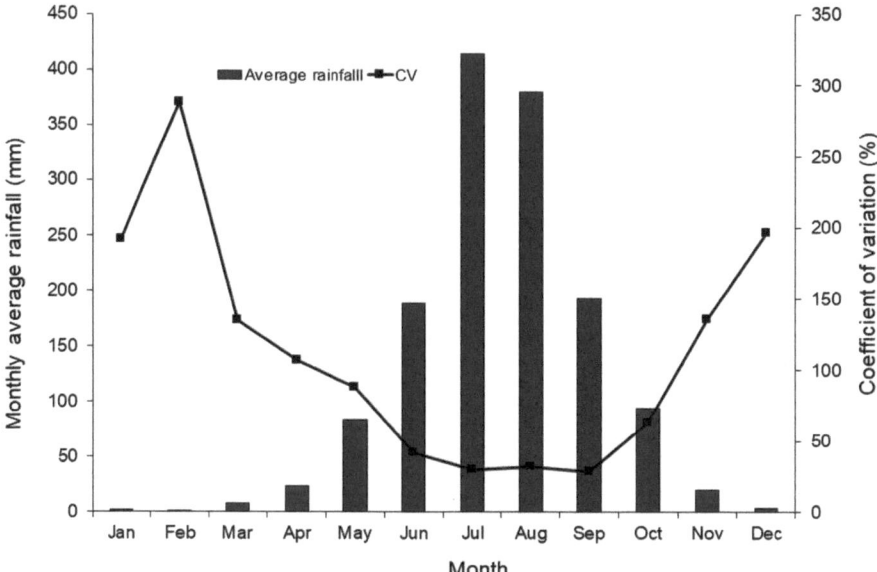

Fig. 4.8 Blue Nile basin mean monthly rainfall and coefficient of variation

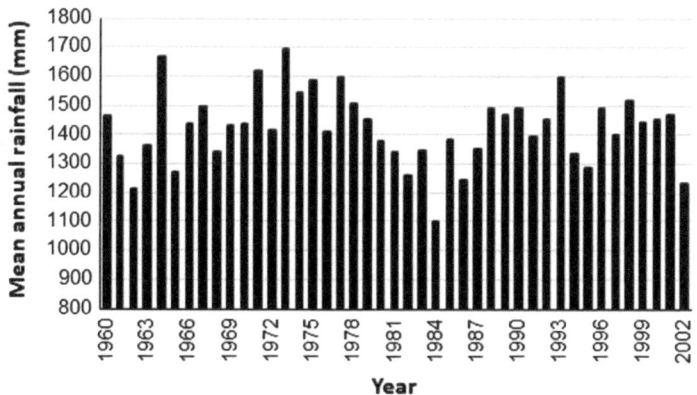

Fig. 4.9 Blue Nile basin average annual rainfall variation (1960–2002)

Niño years. The 2015 strong El Niño and the Ethiopian drought and food shortage is further proof of the link. Major ENSO events are depicted in Table 4.3. A hydrologic modelling of the irrigation and hydropower potential of the Blue Nile basin under climate variability and change showed that doubling the frequency of El Niño will result in reduction of benefit-cost ratio (Block 2007). There were consecutive years with El Niño; 1877–1878, 1899–1900, 1940–1941, 1967–1968, 1982–1983, increasing the impact of drought and famine.

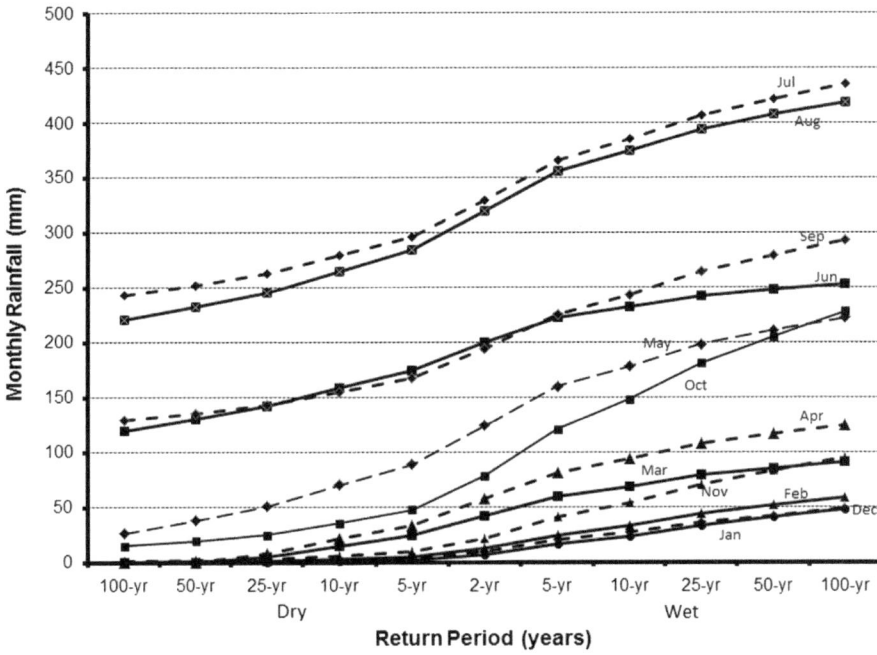

Fig. 4.10 Monthly rainfall for dry and wet return periods

Table 4.3 Strong ENSO events and Ethiopian droughts (bold)

Strong El Niño Years				Strong La Niña Years			
2015	1900	1896	1993	1890	1971	1886	1945
1877	1940	**1969**	1958	1893	1916	1875	1933
1987	1902	1878	1885	1955	1910	1984	1887
1997	1930	1926	1994	1999	1894	2008	1954
1941	1992	1977	2004	1874	1909	1872	2007
1982	1919	1914	1963	1950	2000	1985	1898
1905	**1965**	1957	1915	1975	1956	1873	1964
1888	1991	**1931**	1953	1988	1892	1879	2010
1972	2002	**1983**	2009	1974	1989	1973	2011

4.4 Tributaries of the Blue Nile River

The Blue Nile flows out of Lake Tana. Megech, Ribb, Gumera and Gilgel Abay flow into Lake Tana. The major tributaries into the main stem from the four directions are Beshlo, Woleqa, Jemma, Muger, Guder, Chemoga, Wenchit, Finchaa, Dedessa, Angar, Dura, Rahad, Dinder, Dabus, Gulla and Beles. Figure 4.11 depict the network of tributaries of the Blue Nile.

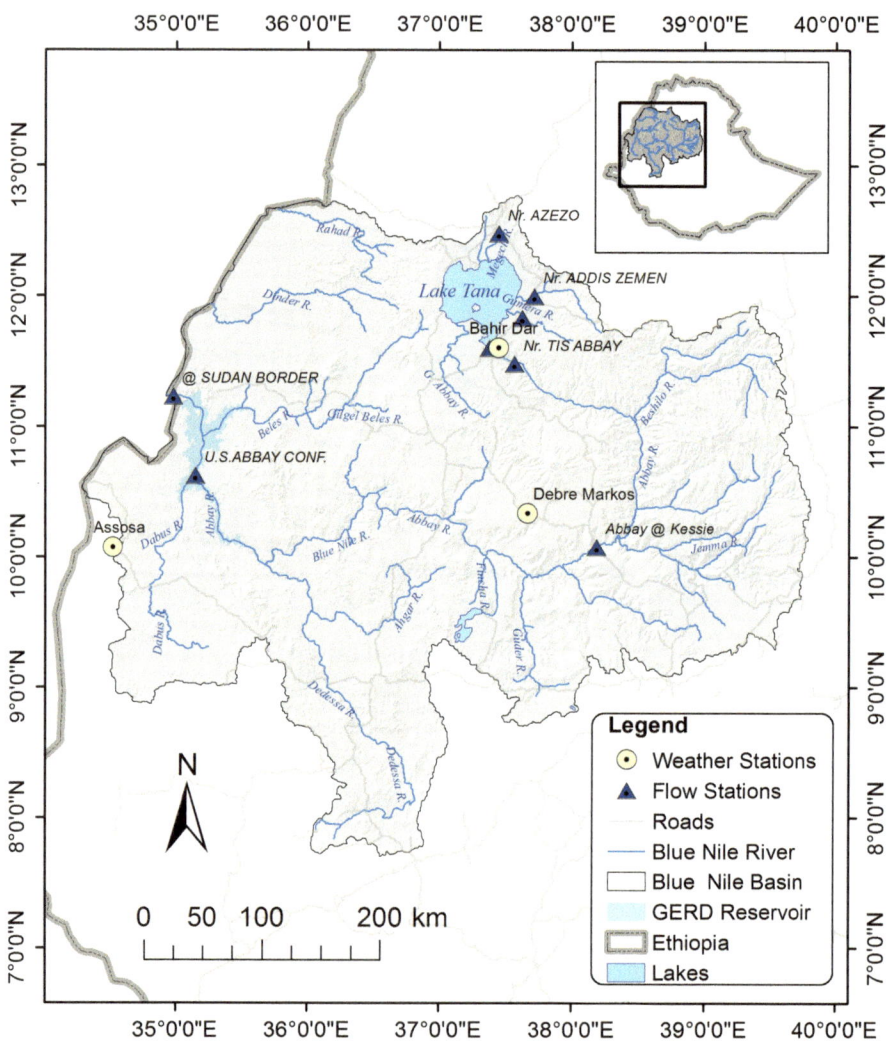

Fig. 4.11 Network of the Blue Nile River tributaries

4.4.1 Lake Tana Watershed

Lake Tana is virtually known as the source of Blue Nile and it constitutes one of the major sub watersheds in the sub basin. Lake Tana subbasin is fed by four major tributaries, Gelgel Abay in the South, Rib and Gummera in the east and Megech in the north (Fig. 4.11). The total watershed area is estimated at 16,000 km^2 with six permanent rivers and 40 seasonal small streams (Ayana et al. 2014). Lake Tana is a highland lake at an average altitude of 1800 m asl, with surface area of 3060 km^2 at an average lake level of 1786 m asl (1964–95). CharaChara weir was commissioned in 2003 at the outlet of Lake Tana with a purpose of regulating the outflow for the generation of hydropower at Tiss Abay, a natural fall located some 30 km downstream of the Lake outlet in the Blue Nile River. After commissioning the CharaChara weir, lake level is observed to increase by about 30 cm. The historic minimum lake level was 1785.13 m (1964–95) observed in June 1983, after the 1982 drought in Ethiopia. The lake has a maximum depth of 15 m (Ayana et al. 2014).

Lake level record for the period of 1964–2003 has been made available for this synthesis work. Maximum lake level occurs in August to September and starts receding in November–December and reaches to minimum level in June.

The annual inflow contributed by Gelgel Abay to the Lake Tana reservoir, is averaged at 1.8 bcm, the Rib and Gummera, at 0.47 and 1.19 bcm respectively and the Megech at 0.23 bcm. In a water balance study, it was reported that annual average rainfall over the lake is 4.68 bcm, gaged flow 2.78 bcm, ungauged flow 4.7 bcm, evaporation 5.87 bcm, outflow 3.73 bcm and change in storage 0.24 bcm (Tegegne et al. 2013). The cited literature reported the water balance parameters in mm year^{-1}. Conversion into volume was made based on lake area of 3626 km^2. Figure 4.12 depict Lake Tana and its drainage area.

Due to limitation in flow data availability, only few flow patterns for few of the tributaries are presented. Figure 4.13 depicts mean monthly flow for the main streams that feed to Lake Tana (Megech, Ribb, Gumera, Gilgel Abay and Koga). Megech, Ribb, Gumera and Gilgel Abay flow directly into the lake while Koga flows into Gilgel Abay. Chemoga is tributary to the Blue Nile. Dry season flows are low while July, August and September have high flows.

Tributaries of the Blue Nile are more accessible for local water abstraction and relief to dependence on rainfed agriculture. To combat periodic famine from drought and poverty, Ethiopia has made some progress to benefit from running water through small scale diversions and irrigation schemes. Planned irrigation projects are reported where some may be complete at this time. Alemayehu et al. (2009) reported planned irrigation in Lake Tana basin of 12,852 ha (Gilgel Abay), 14,000 ha (Gumera), 19,925 ha (Ribb) , 7300 ha (Megech), 6000 (Koga), 5745 ha (NE Lake Tana) and 5132 ha (SW Lake Tana). Other planned water resources developments are irrigation in Beles subbasin, 30,731 ha (Upper Beles) and 85,000 ha (Lower Beles); 12,000 ha from Finchaa dam and 14,450 ha from Anger (McCartney et al. 2009). The Arjo Didessa irrigation project to irrigate 80,000 ha is undergoing with dam and infrastructure construction. The Blue Nile River is fully utilized in Sudan and Egypt for irrigation, hydropower, water supply, fishing and recreation with a potential of cross boundary water sale.

Densaw et al. (2016) reported Koga dam is completed and provide irrigation water for small farmers. They also stated that 11 similar efforts are planned to expand irrigation in the Tana sub-basin of the Blue Nile basin. Ribb reservoir is under construction. Twenty thousand ha irrigation plan is being implemented on Megech, Ribb and Anger rivers with pumps and dams (Ministry of Water Resources of Ethiopia 2010). The Tana-Beles project is a 12 km tunnel and diversion to the southwest for irrigation and power generation.

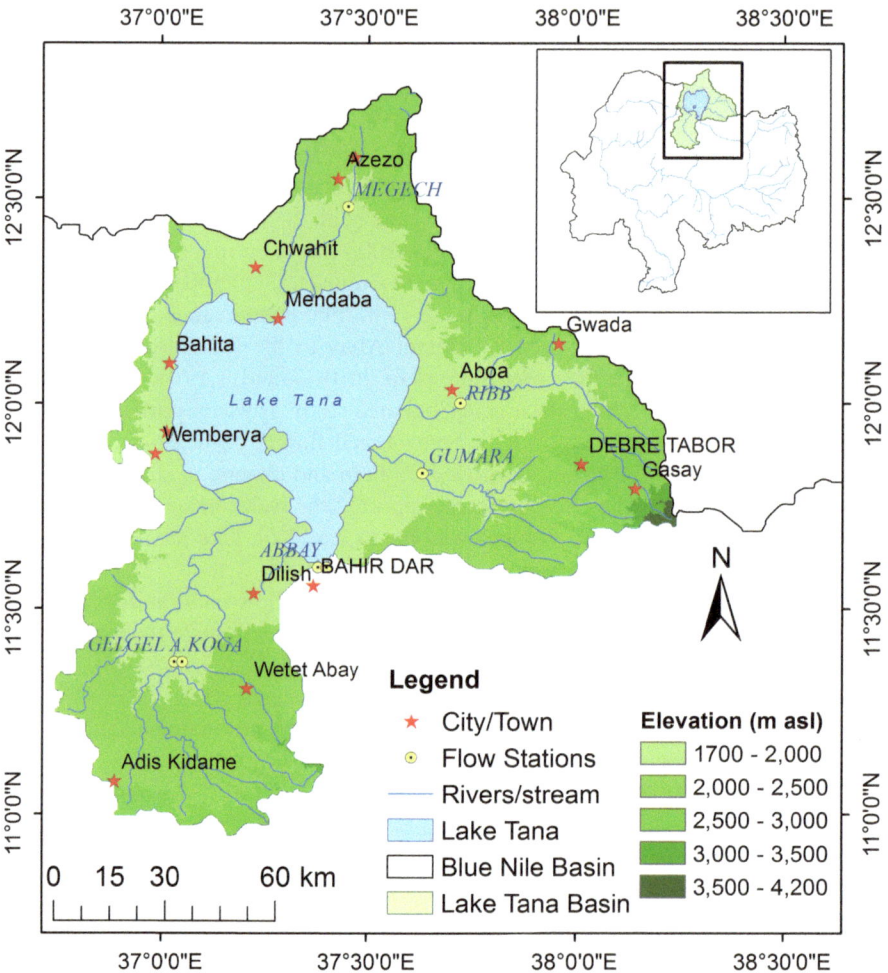

Fig. 4.12 Lake Tana and drainage area

4.4.2   Abay at Kessie

This station is one of the major gauging stations in the main stem of the Blue Nile. With a total area of 65,784 km^2 it covers about 40% of the Abay area at the border and nearly 20% of the total sub basin. It is located downstream the Addis-Bihar Dar Highway Bridge that crosses the main stem of the Blue Nile. The mean annual inflow passing Kessie is averaged at 17.42 bcm that accounts about 35% of the Blue Nile mean annual flow at the border. Flow at Kessie has seasonal variability with more than 85% of the runoff concentrated in the months of June to October, with the peaks in August.

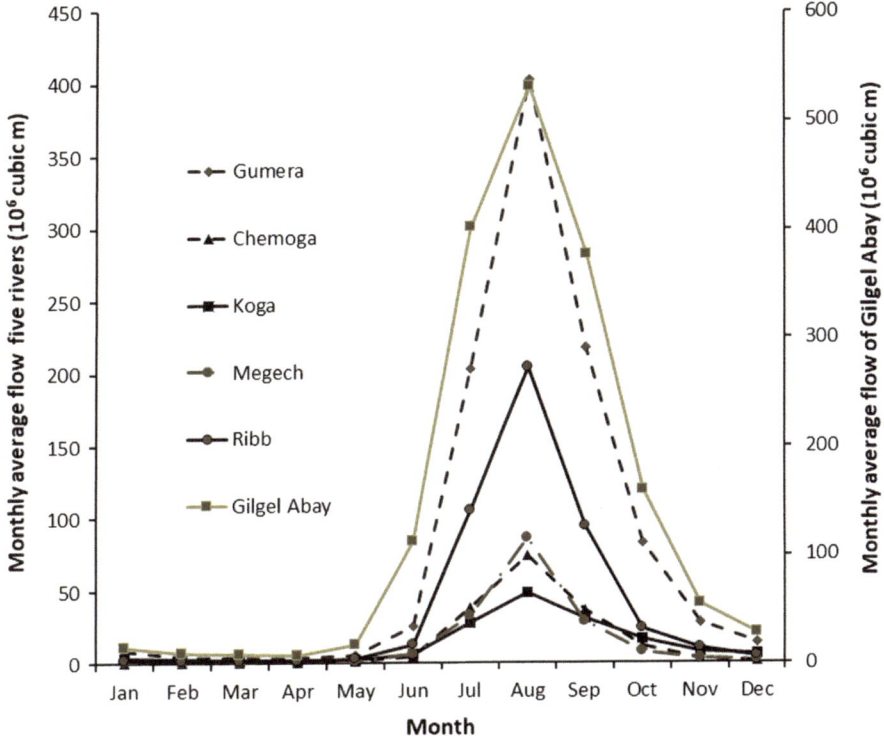

Fig. 4.13 Blue Nile tributary rivers mean monthly flow and seasonal distribution (1959–2002)

Dedessa, Dabus and Angar rivers confluence with the main Blue Nile River downstream of the Kessie station in its left bank. The mean annual contribution of Dedessa River is averaged at 4.6 bcm, the Angar at 1.8 bcm and that of the Dabus at 3.2 bcm.

4.4.3 Beles River

The Beles River has two major tributaries, upper Beles and the Gelgel Beles. At the mouth, the mean annual contribution of the Beles River to the main Blue Nile is averaged at 1.6 bcm. Beles is the last major tributary of the Blue Nile before crossing the Ethio-Sudan border.

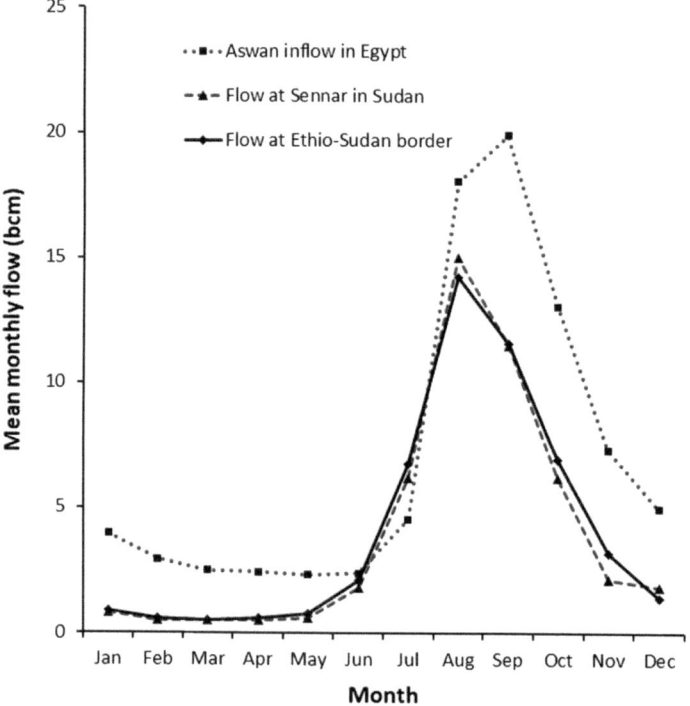

Fig. 4.14 Mean monthly Nile flow with seasonal variation (*Data source* Sadek (2006), Mulat et al. (2014)

4.4.4 Blue Nile at the Ethio-Sudan Border

The inflow of Blue Nile at the border has been measured since 1960. The long-term average (1960–2003, Abay Master Plan Studies, Feb 1998) indicates that mean annual inflow of the Blue Nile at the border is averaged at 51.30 bcm. The short term (1999–2003) mean inflow is averaged at 54 bcm.

Nile river flow seasonal distribution varies with rainfall distribution in both the Blue and White Nile basins and flood travel time. Figure 4.14 shows mean monthly flow seasonal variation of the Blue Nile River at Ethio-Sudan border (1967–1972; Mulat et al. (2014); 1999–2003); inflow into Sennar reservoir (Mulat et al. 2014) in Sudan and inflow into Aswan Reservoir (Sadek 2006). Average annual flow at Ethio-Sudan border is 49.5 bcm; Sennar is 47.6 bcm and Aswan is 84.31 bcm for the above cited data. Flow temporal variation patterns rainfall variation except river flow lags with distance from the runoff generating basin. Flood flow at Ethio-Sudan border and Sennar dam is from July to October while it is August to November at Aswan (Fig. 4.14). Ethio-Sudan border and Sennar flows are from the Blue Nile while Aswan receives additional flows from the White Nile, Atbara and Sobat (Fig. 4.15).

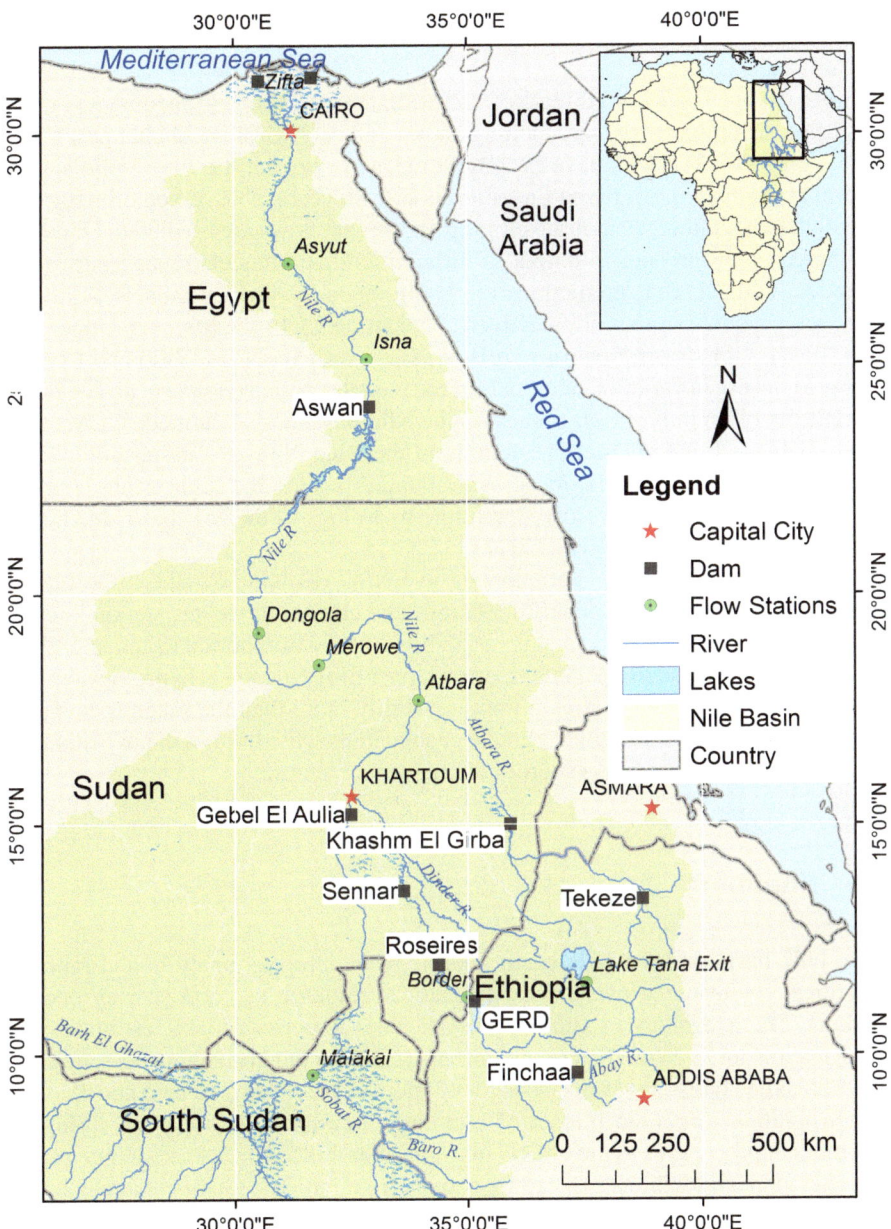

Fig. 4.15 Dams on the Nile River

4.5   Climate Change and the Nile Basin

Climate change is a global threat with immediate impacts being documented on sea level rise, rising temperatures, melting snows and ecohydrological changes. Global Climate Models (GCMs) have become a means to predict the future climate of a region as the Nile basin based on calibration from years of observed climatic data. Rainfall, temperature, evapotranspiration and stream flows are predicted. One of the challenges is differences in output of different climatic models raising uncertainty in model outputs. But, predictions are being made by various models as average of several model outputs or selectively. Validation of 14 GCMs over Upper Blue Nile Basin with stream flow of two Blue Nile river tributaries, Ribb and Gumera, showed concern in variation of outputs of the models applied (Bokke et al. 2017). The complexity of rainfall run-off process in the Nile Basin has so far made GCM results less reliable. Climate model projections in the Blue Nile are further complicated due to rugged terrain forcing orographic rainfall pattern and high local variability. The problems of GCM to resolve the future hydrology of the Nile basin and horn of Africa is discussed by Conway (2017).

Linking climate change with change in ENSO (El Nino Southern Oscillation) pattern and based on empirical observations and climate model projection, increase in Nile flow (15%) and interannual variability (50%) was reported (Siam and Eltahir 2017). Studies of modelling application on Upper Blue Nile Basin hydrology reported that ENSO events change stream flow variability more than the magnitude (Elsanabary and Gan 2015). Regional climate change in the Nile basin will have impact on Nile basin water and other resources in one way or another.

4.6   Summary

Blue Nile basin and the Nile basin hydrology monitoring, analysis and reporting is critical for resource availability evaluation, resource use and sharing decision making. The hydrology of Blue Nile is the basis to derive reliable estimate of the available resource. With the completion of GERD on the horizon, understanding of the Blue Nile Hydrology is the basis for efficient utilization of the dam and resolving water conflict. Since the impact of hydrology variability is apparent, filling and operational procedures to the GERD may consider this factor. It is also imperative to conduct more simulations assisted by outputs from multiple GCMs that may be able to reproduce the historical climate pattern of the region and produce prediction with some reliability. Downstream flow variation due to hydrometeorology variability may be mistaken for upstream dam operations.

References

Abtew W, Melesse AM, Dessalegne T (2009a) El Niño southern oscillation link to the Blue Nile River hydrology. Hydrol Process 23:3653–3660

Abtew W, Melesse AM, Dessalegne T (2009b) Spatial, inter and intra-annual variability of the Upper Blue Nile Basin rainfall. Hydrol Process 23:3075–3082

Abtew W, Melesse AM (2013) Evaporation and evapotranspiration measurements and estimations. Springer, New York

Alemayehu T, McCartney M, Kebede S (2009) Simulation of water resource development and environmental flows in the Lake Tana subbasin. In: Aulachew SB, Erkossa T, Smakhatin V, Fernando A (eds) Improved water and land management in the Ethiopian highlands: its impact on downstream stakeholders dependent on the Blue Nile. IWMI. Addis Ababa, Ethiopia

Ayana EK, Philpot WD, Melesse AM, Steenhuis TS (2014) Chapter 14 Bathymetry, lake area and volume mapping: a remote-sensing perspective. In: Melesse AM, Abtew W, Setegn (eds) Nile River Basin ecohydrological challenges, climate change and hydropolitics. Springer, New York

Berhanu B, Sileshi Y, Melesse AM (2014) Chapter 6 Surface water and groundwater resources of Ethiopia: potentials and challenges of water resources development. In: Melesse AM, Abtew W, Setegn (eds) Nile River Basin ecohydrological challenges, climate change and hydropolitics. Springer, New York

Block PJ (2007) Integrated management of the Blue Nile basin in Ethiopia hydropower and irrigation. IFPRI, Washington, DC

Bokke AS, Taye MT, Willems P, Siyum SA (2017) Validation of general climate models (GCMs) over Upper Blue Nile Basin, Ethiopia. Atmos Clim Sci 7:65–75

Conway D (2017) Future Nile River flows. Advance Online Publication. www.nature.com/naturec limatechange

Degefu GT (2003) The Nile historical legal and developmental perspectives. Trafford Publishing, Victoria

Densaw DF, Ayana EK, Enku T (2016) Koga irrigation scheme water quality assessment, relation to streamflow and implication on crop yield. In: Melesse AM, Abtew W (eds) Landscape dynamics, soils and hydrological processes in varied climates. Springer, New York

Dessu SB, Melesse AM (2013) Impact and uncertainties of climate change on the hydrology of the Mara River Basin. Kenya/Tanzania Hydrological Processes: (in press). https://doi.org/10.1002/h yp.9434

Dessu SB, Seid AH, Abiy AZ, Melesse AM (2016) Chapter 18: Flood forecasting and stream flow simulation of the Upper Awash River basin, Ethiopia, using geospatial stream flow model (GeoSFM). In: Melesse AM, Abtew W (eds) Landscape dynamics, soils and hydrological processes in varied climates. Springer, New York

Elsanabary MH, Gan TY (2015) Evaluation of climate anomalies impacts on the Upper Blue Nile Basin in Ethiopia using distributed and a lumped hydrologic model. J Hydrol 530:225–240

ESA (2008) GlobCover land cover v2 2008 database. European Space Agency. European Space Agency GlobCover Project, led by MEDIAS-France. http://ionia1.esrin.esa.int/index.asp

FAO, IIASA, ISRIC, ISSCAS, JRC (2009) Harmonized world soil database (version 1.1)

Fischer G, Nachtergaele F, Prieler S, van Velthuizen HT, Verelst L, Wiberg D (2008) Global agro-ecological zones assessment for agriculture (GAEZ 2008)

Gissila T, Black E, Grimes DIF, Slingo JM (2004) Seasonal forecasting of the Ethiopian summer rains. Int J Climatol 24:1345–1358

Glantz MH (ed) (1987) Drought and hunger in Africa: denying famine a future. Cambridge University Press, Cambridge

Halfman JD, Johnson TC, Finney BP (1994) New AMS dates, stratigraphic correlations and decadal climatic cycles for the past 4 ka at Lake Turkana, Kenya. Palaeogeogr Palaeoclimatol Plaeoecol 111(1–2):83–98

IPRC (2011) Tree rings tell 1100-year history of El Niño. Pres Release. The School of Ocean and Earth Science and Technology at the University of Hawaii at Manoa

Jarvis A, Reuter HI, Nelson A, Guevera E (2008) Hole-filled SRTM for the globe Version 4. CGIAR—consortium for spatial information. http://srtm.csi.cgiar.org/

McCartney M, Ibrahim YA, Sileshi Y, Awulachew SB (2009) Application of the water evaluation and planning (WEAP) model to simulate current and future water demand in the Blue Nile. In: Aulachew SB, Erkossa T, Smakhatin V, Fernando A (eds) Improved water and land management in the Ethiopian highlands: its impact on downstream stakeholders dependent on the Blue Nile. IWMI. Addis Ababa, Ethiopia

Ministry of Water Resources of Ethiopia (2010) Environmental and social impact of about 20,000 ha irrigation and drainage schemes at Megech pump (Seraba), Ribb and Anger dam. Report 60386 V1

MoWR (2016) Ministry of water resources of Ethiopia. http://www.mowr.gov.et/index.php?pagenum=2.3&pagehgt=5395px. Accessed 13 Nov 2016

Mulat AG, Moges SA, Ibrahim Y (2014) Chapter 27 Impact and benefit study of Grand Ethiopian Renaissance Dam (GERD) during impounding and operation phases on downstream structures in the Eastern Nile. In: Melesse AM, Abtew W, Setegn (eds) Nile River Basin ecohydrological challenges, climate change and hydropolitics. Springer, New York

Pankhurst R (1966) The great Ethiopian famine of 1888–1892: a new assessment part one. J Hist Med Allied Sci, XXI:95–124

Putter TD, Loutre MF, Wansard G (1998) Decadal periodicities of Nile River historical discharge (A.D. 622–1470) and climatic implications. Geophys Res Lett 25(16):3193–3196

Sadek N (2006) River Nile flood forecasting and its effect on national projects implementation. In: Tenth international water technology conference, IWTC102006, Alexanderia, Egypt

Seleshi Y, Zanke U (2004) Recent changes in rainfall and rainy days in Ethiopia. Int J Climatol 24:973–983

Semazzi FHM, Burns B, Lin N, Schemm J (1996) A GCM study of the teleconnections between the continental climate of Africa and global sea surface temperature anomalies. J Clim 9(1):2480–2497

Siam MS, Eltahir EAB (2017) Climatic change enhances interannual variability of the Nile river flow. Nat Clim Change 7(5):350–354

Tegegne G, Hailu D, Aranganathan SM (2013) Lake Tana reservoir water balance model. Inter J Appl Innov Eng Mang 2(3):474–478

Chapter 5
Grand Ethiopian Renaissance Dam Site Importance

Abstract The Grand Ethiopian Renaissance Dam (GERD) site selection is influenced by dam engineering, political objectives, economic benefits, than dam security, environmental and social dynamics. The GERD site is about 20 km from the Ethio-Sudan boarder in the Benshangul-Gumuz region of Western Ethiopia. This chapter evaluates the hydrological and geological or geotechnical characteristics of catchment and dam site with respect to suitability of the GERD reservoir site. Detailed hydrological review to the Blue Nile basin is covered in Chap. 4. The influence of domestic and regional socio-political, economic and environmental factors on GERD site selection were reviewed.

Keywords Grand Ethiopian Renaissance Dam · Ethiopia · Egypt · Sudan Transboundary Rivers · Nile basin · Benshangul-Gumuz · Ethnic mapping

5.1 Introduction

Large dams have political, economic, environmental and social impacts. In case of transboundary rivers water rights issues with riparian countries becomes a source of potential conflict. Big dam projects have mostly unintended consequence such as environmental degradation, displacement of settlements with in the flooded area of the reservoir along with economic and cultural losses. The environmental consequences of large dams are paramount from inundation of faunal and flora by the reservoir, altered hydrologic regimes at downstream, ecological impacts, and depriving essential nutrients to deltas and coastal estuaries. The national importance of big dams and socio-cultural and environmental impacts are reported including GERD (Veilleux 2013, 2014). Ideally, building a dam worth a good percentage of the GDP of a country requires the utmost responsibility to research the economic, environmental, security and social benefits of the project. This requires the participation of intellectuals, political parties and other social groups to determine the need, the location and funding and to secure sustainable public support and commit to defend in case of attack by adversaries. In the case of the GERD on the Blue Nile, the clandestine planning of the dam is understandable from potential opposition from riparian countries. But, the clandestine planning and decision was also interpreted by internal political opposition groups as purposeful attempt by the ruling party to hide the dam site selection and obscure the real objective from the public.

5.2 Major Dams in the Nile River Basin

There are over 25 dams and major water control infrastructures on the Nile River (Fig. 5.1). GERD is the first major dam to be built on the main Blue Nile River of Ethiopia. Sennar and Roseires are the two major dams on the Sudan on the Blue Nile River. The Blue Nile basin covers 10% of the Nile Basin but contributes 60% of the flow to the Main Nile River in Sudan and Egypt. The High Aswan Dam (HAD) in Egypt is the largest dam in the Nile River Basin with a reservoir capacity of 132 billion cubic meter (bcm). The HAD has been perceived as a political symbol and achievement (Van Der Schalie 1974). The GERD is the first and only dam being built on the Blue Nile river of Ethiopia. But, water rights issues have generated downstream resistance to the GERD. The major factor for downstream concern is potential flow reduction and lack of control of the dam operation. Looking at the regional populations and economic growth, the number of existing dams on the Nile will likely increase in the coming decades to meet the growing energy and water demand in the basin and GERD is just one of them.

5.3 Review of GERD Dam and Reservoir Features

During the construction of the Aswan High Dam in Egypt with the support of the Soviet Union, the United States Bureau of Reclamation (USBR) studied potential hydropower dam sites on the Blue Nile in Ethiopia from 1956 to 1964. In 1964, it offered four proposed dam sites, Karadobi, Mabil, Mendia and Border (Fig. 5.2). The current Ethiopian Government updated USBR studies with French and later Dutch consultants in 1998 (Saeed 2018).

These sites were promoted as co-operative development of the Nile basin by Nile Basin Initiative, Eastern Nile Technical Regional Office. The location of the dam has certain advantages and disadvantages given the stated purpose of the dam is to generate power and export power to neighbouring Sudan and Egypt. The dam location near the Sudan lowers cost of transmission lines. The site also may have offered construction ease with respect to the amount of work needed for diversion and dewatering during the construction of the dam, compared to alternative sites upstream. But, disadvantages include that large potentially irrigable and mineable land will be permanently inundated as a result of site selection on relatively flat terrain that transitions from the Blue Nile gorge to the lowlands (Fig. 5.3). The land size under the dam at peak water level of 640 m asl is over 1700 km^2. Seepage

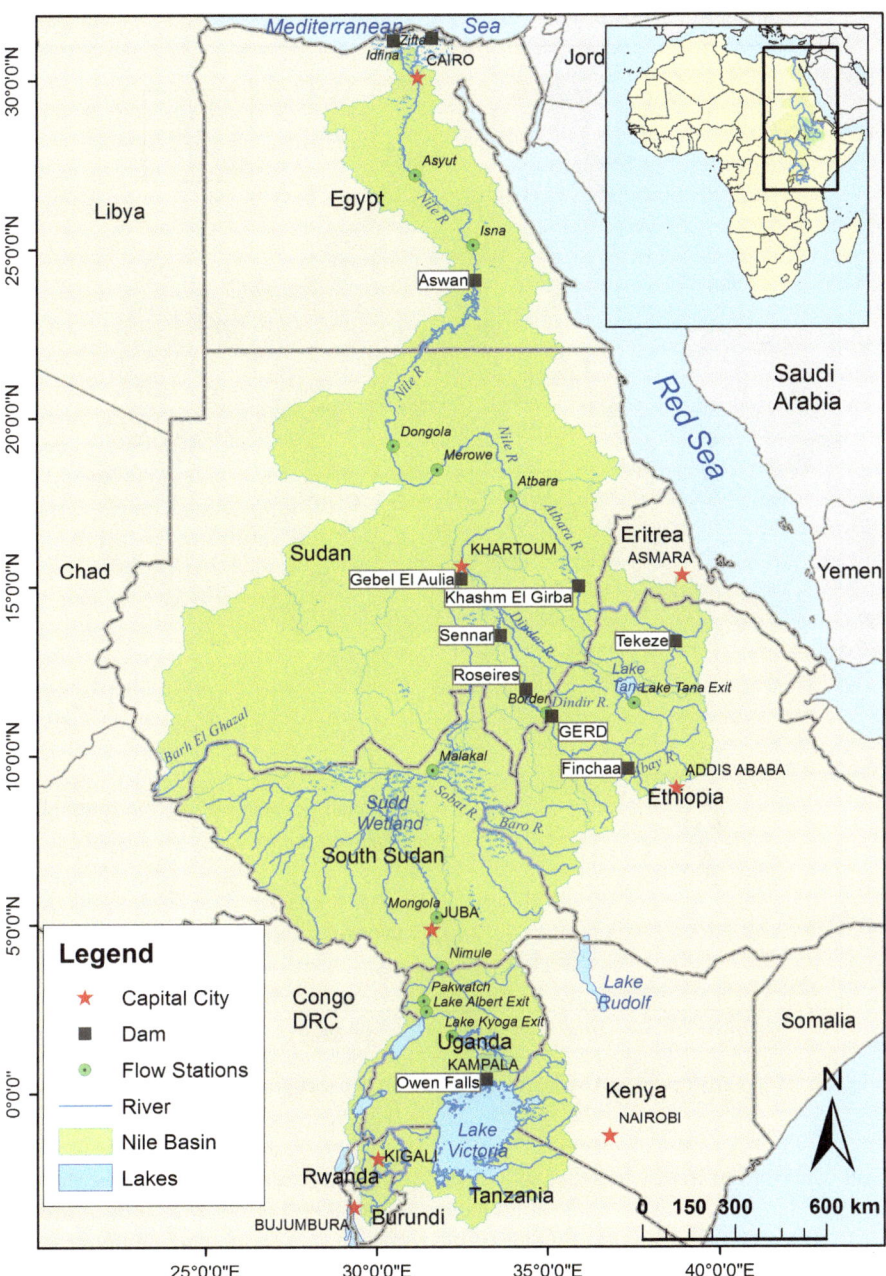

Fig. 5.1 The Nile basin and location of major dams

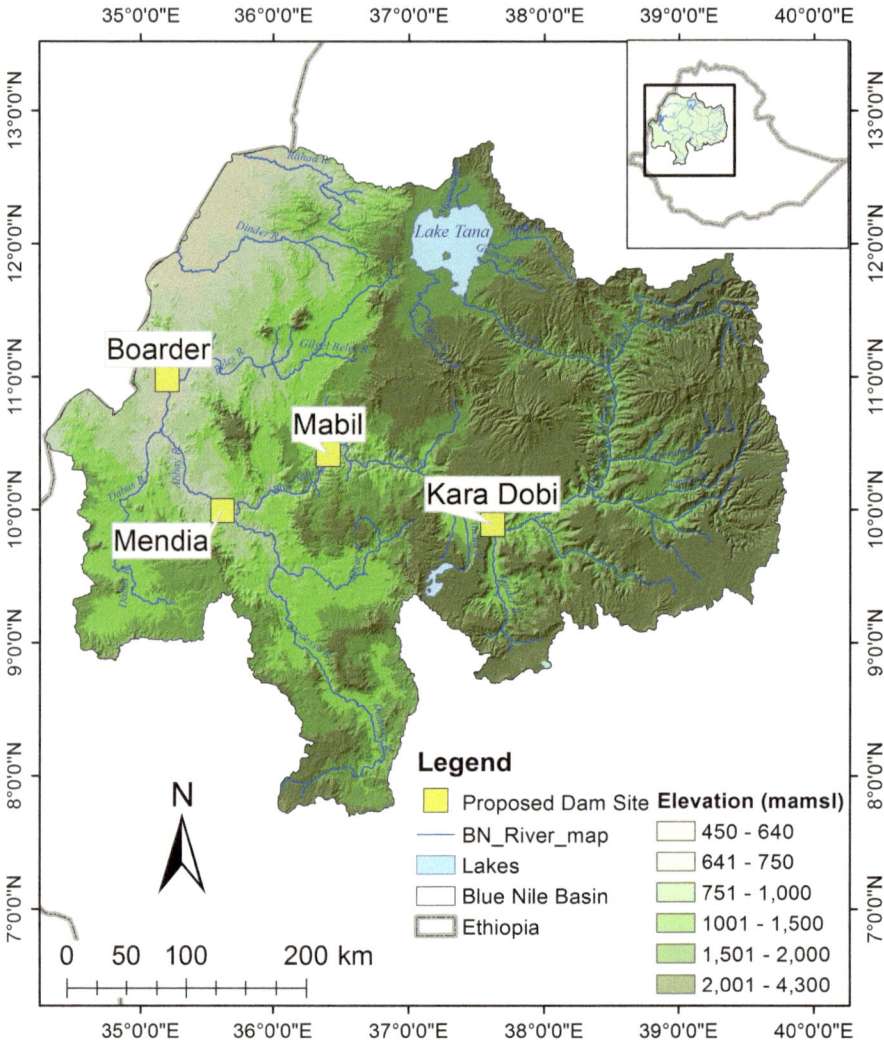

Fig. 5.2 1964 USBR proposed hydroelectric dam sites and topography of the Blue Nile basin

and evaporation losses increase as a result of the large surface area of the dam and overall dam construction and maintenance cost will be high. The length of the dam is 1800 m. The increased perimeter of the reservoir increases seepage potential and its management. The dam site and associated businesses has the potential of larger

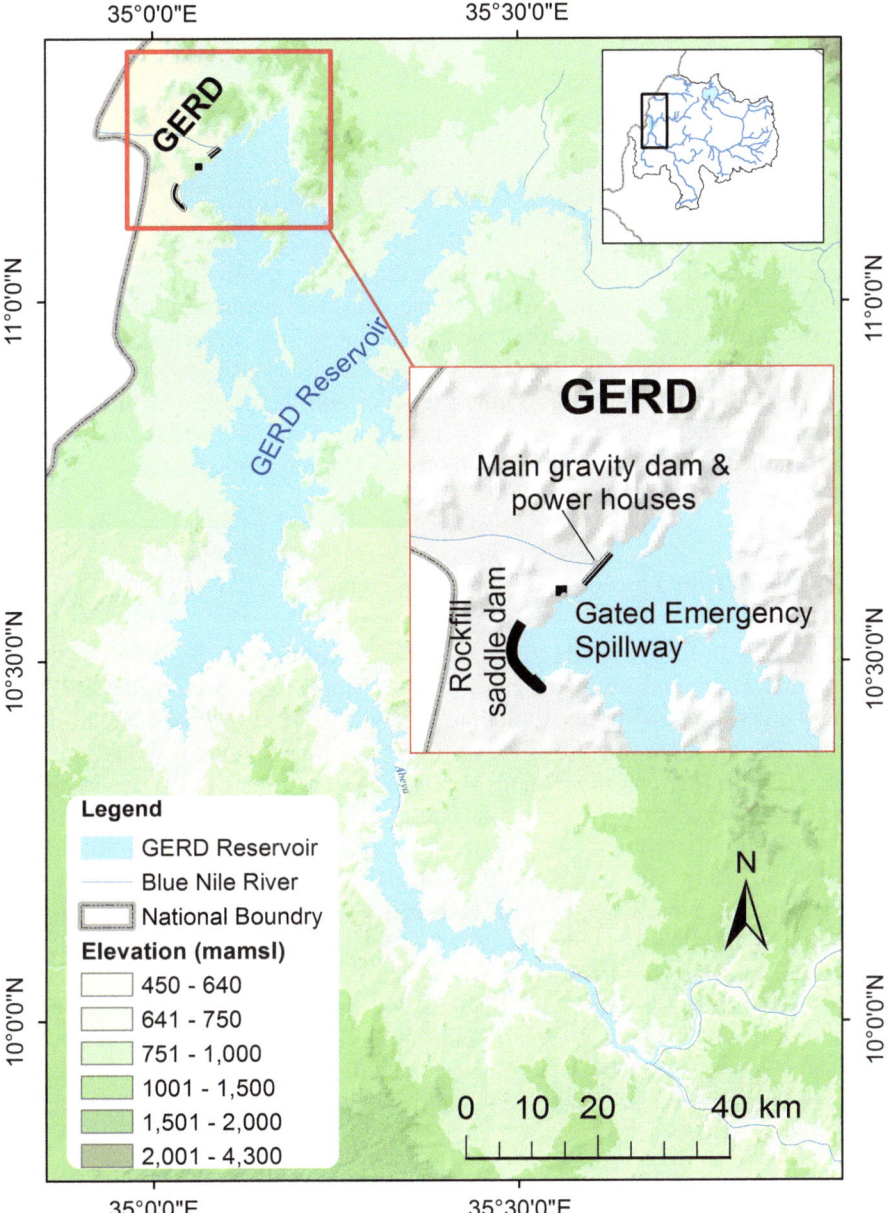

Fig. 5.3 GERD site in Benshangul Gumuz region about 20 km from Ethiopia/Sudan border

displacement of the ethnic people of the area with powerful ethnic groups as is the case in the capital city, Addis Ababa and the agricultural lands of Gambela (Legesse 2014; Pearce 2012).

GERD is a combination of main gravity dam, spillway and rockfill dam stretched over more than 6 km dam axis. The GERD reservoir stretches more than 246 km upstream of the dam.

5.4 Technical Features of GERD Site Selection

Technical acceptability of the GERD site is dictated by the hydrology of the catchment, presence of a satisfactory site for the dam, the availability of construction materials suitable for dam construction, and the integrity of the reservoir basin with respect to leakage. Functional suitability of GERD site is also governed by the balance between its natural physical characteristics and the purpose of the reservoir. Functional and technical requirements for selecting a site for the dam and reservoir were compiled and reviewed here based on data from multiple sources.

5.4.1 Hydrology

Catchment hydrology, available head and storage volume etc. must be matched to operational parameters set by the nature and scale of the project served. GERD is at the most downstream end of the Ethiopian Blue Nile River just before the Ethio-Sudan boarder. The location maximizes the utilization of flow volume from all the major tributaries starting from Ethiopian highlands to be stored in the GERD reservoir. The Blue Nile basin covers 199,812 km^2 surface area. The drainage area for contributing flow to GERD reservoir is 172,250 km^2. The annual flow of the Blue Nile is 54.4 (bcm), with 746 m^3 d^{-1} km^{-2} runoff generation rate as reported by Berhanu et al. (2014). Annual rainfall ranges from less than a 900 mm in the Northeast to over 2000 mm in the south (Abtew et al. 2009). On the basis of annual rainfall and river discharge, the Blue Nile basin is the wettest part of Ethiopia.

The tributaries of the Blue Nile River are Woleqa, Beshlo, Jemma, Muger, Guder, Chemoga, Fincha, Didessa, Beles, Angar, Dabus, Gilgel Abay, Ribb, Gumera, Wenchit and numerous streams. Runoff is generated from urban, agricultural, forest and other undevelopable areas. Figure 5.4 depicts tributaries of the Blue Nile River.

5.4.2 Topography

Considering Ethiopia's intention to export power to Sudan and Egypt, the dam site is at the nearest possible location to reduce transmission grid cost, and minimize power conveyance losses (Fig. 5.1). The topography of GERD site has provided minimal cost of stream diversion and dewatering with a combination of construction schedule from abutments to the natural river course and subsequent provision of four bottom

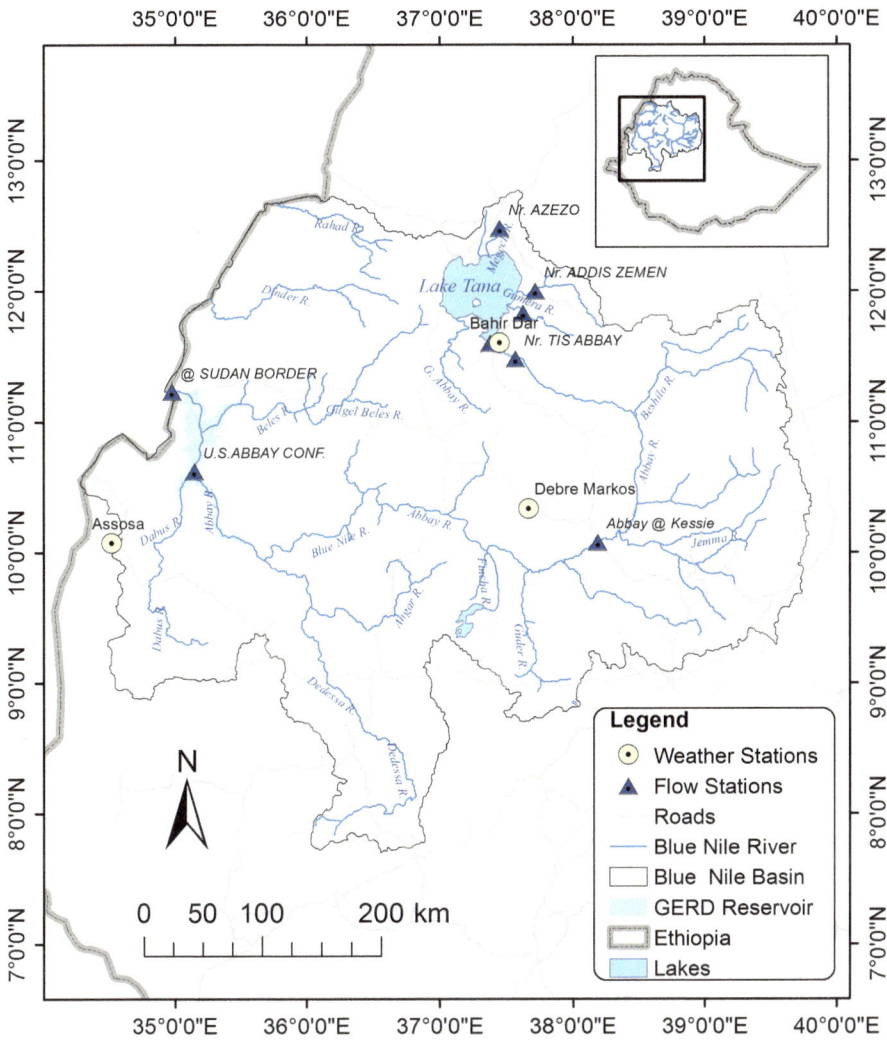

Fig. 5.4 Tributaries of the Blue Nile

diversion outlets. The three hills along the dam axis enabled the required storage and hydraulic head with combination of gravity and saddle dam, with a separate emergency. Two spillways were included; one as part of the main gravity dam and the other as a separate emergency spillway between the two dams to ensure safety of the rock-fill saddle dam. The emergency spillway diverts floodwater back to the natural stream downstream of the main gravity dam.

5.4.3 Geology

GERD is a combination of 170 m high gravity dam, 50 m high rockfill dam and a separate gated spillway. The combined length of the dams and spillways is more than 6 km. The length and weight of the dams may force crossing multiple geologic formations and features with varying structural integrity. Since 70% of the dam has been built at the beginning of 2018, only remediation work can be done to reduce any risk associated with geologic features of the site. Due to lack of detailed geologic study at the dam site, a review of geologic features based on a small scale geological map (1: 2,000,000) is presented here (GSE 1996). The GERD site is dominated by pre-Cambrian metamorphic rocks (Fig. 5.5) (Tadesse et al. 2003). These formations likely provide sufficient foundation strength to support the main gravity dam with minimal cost of construction and maintenance.

The GERD site is dominated by pre-Cambrian metamorphic rock formation. The geology provides strong foundation to support the heavy weight of the main gravity dam and necessary rock material for the rockfill dam. There is possible fault line of a 2–3 km possibly stretching across the dam axis. Note: Clay and silt soils overlay the GERD dam site. Given the small scale of the geologic map, the exact interface of the formations and features may not be accurately related to GERD.

5.5 Environmental, Social, Economic and Political Considerations

Ethiopia's unilateral decision to construct GERD on the Blue Nile (Abay) River was not only a surprise to the regional power structure but also elevated the longstanding environmental, economic and socio-political issues associated with development of Nile waters. GERD draws parallel with the High Aswan dam in Egypt. The environmental impacts and other socio-political considerations of GERD extends across a diverse spectrum of issues from population grown, economic development, water right; sedimentation and/or of changing flood regime and climate change. It is necessary to examine the complex social and environmental values of water resources and the policies governing the use of the resource.

5.5.1 Environmental Impacts of GERD Site

Technical suitability of the GERD site evaluation must be augmented with assessment of the anticipated social and environmental consequences of construction and operation of the dam. Mitigation planning can be done. The high sediment yield from the traditional farming and high slope will likely affect the storage capacity and life-span of the GERD. The GERD will alleviate most of the sedimentation prob-

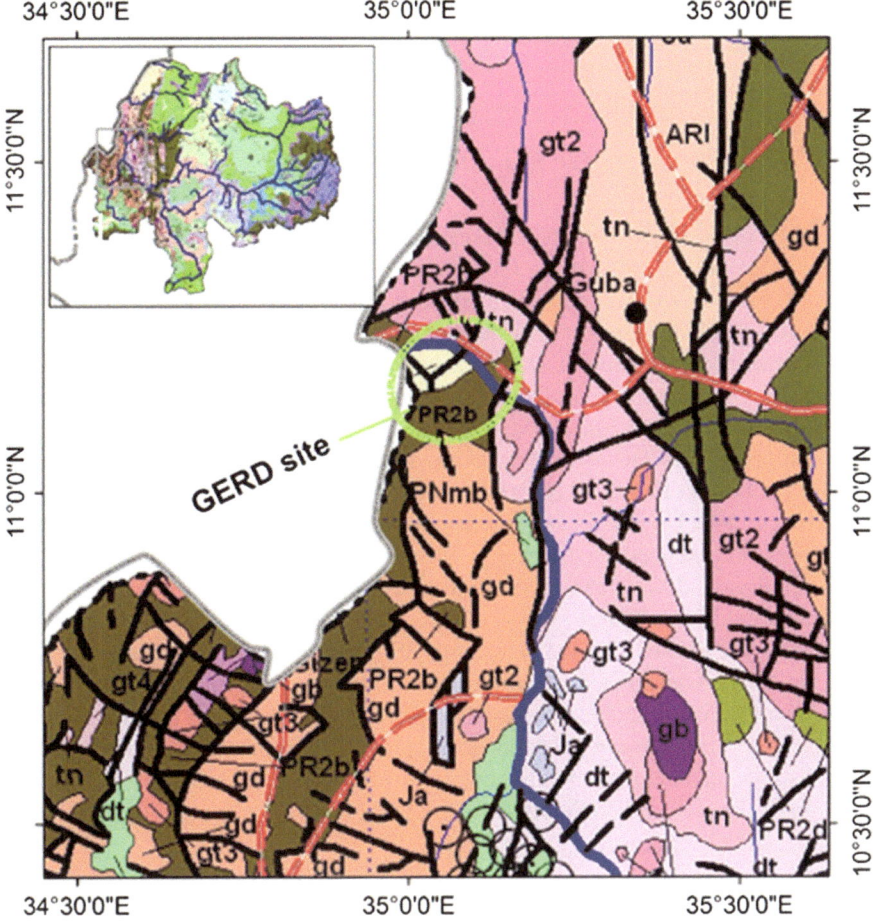

Geologic Formations at the GERD site

Q	Sand, silt, clay, diatomite, limestone and beach sand.
PR2b	Metabasalt, metaandesite, metarhyolite, phyllite, graphiticshchist, margle, quartzite, metaconglomerate, greenshist, metasandstone, metachert and amphibolite
gt2	Syn - tectonic granite
tn	Tonalite

Fig. 5.5 Geologic map of the Blue Nile Basin/GERD site (GSE 1996)

lems of downstream reservoirs. Annually, 207 million m^3 is of sediment is expected at GERD (IPoE 2013). The Blue Nile watershed management through sustainable agriculture, afforestation and economic growth in the basin is in the interest of both upstream and downstream users.

5.5.2 The People at GERD Site

Compared to the upper Blue Nile basin, The GERD dam site is sparsely populated by the indigenous people of the Gumuz and Berta and others who have moved in from other ethnic zones. Approximately 20,000 people are affected by the dam who are mostly subsistence farmers, fishers and hunters (Veilleux 2013). Relatively, the reservoir area will have minimal loss of submerged property except a bridge upstream. This is not to minimize the settlement and way of life loss of the native people. There is practically no relocation of major roads except a bridge, buildings or any other major infrastructure. However, the southwestern region of the reservoir is rich in gold and iron minerals (Tadesse et al. 2003). Panning for gold is a source of income for the local inhabitants. Livestock, hunting, honey collection, handicrafts, charcoaling and trade are also practiced by the people. Details of the dam site including population statistics flora and fauna are reported in a field visit report by International Rivers (International Rivers 2012).

Development associated with the dam will change the ethnic makeup of the region with potential displacements and conflicts. Urban development, tourism, fishery, water front business and residences and other opportunities will result making few rich with the same framework of "economic growth" as in other parts of the country as demonstrated in the capital, Addis Ababa with urban land and in Gambela agricultural land grabs (Legesse 2014; Pearce 2012).

5.5.3 GERD Site and Regional Hydro-politics

The Ethiopian government selected the Boarder site and began construction of the GERD in April 2011 concealing the plan from the public first naming Project X, then Millennium Dam and finally Renaissance Dam. The Dam was announced on March 12, 2011; contract was signed with Salini next day and cornerstone placed on April 2, 2011 (Saeed 2018). The sole purpose of the dam was announced to be hydroelectric power for consumption in Ethiopia and export to Egypt and Sudan. The launching of the GERD project surprised northeaster Africa (Egypt), (Gebreluel 2014). It grabbed the attention of riparian countries and international experts. Considering Ethiopia's intention to export power to Sudan and Egypt, the dam site is at the nearest possible location to reduce transmission grid cost, and minimize power conveyance losses.

The Ethiopian government argues the GERD will not reduce the annual flow volume. The location of the dam being far downstream from potential irrigable area as well as any other potential consumptive water use is presented as Ethiopia's intent for co-operation with downstream countries on the Nile waters development.

However, with the potential of water conflict, its proximity to a neighbouring country (20 km) will put it at a high risk of sabotage, interference, and in extreme case, loss of control of the dam during conflicts.

5.5.4 GERD Site and Ethiopia Domestic Politics

According to Abdelhady et al. (2015), the dam is to create a sense of unity or nationalism in contrast to "difference of politics" of the country's ethnic federalism. The site of the dam has great implication with the current political system of Ethiopia. The current political system and constitution of Ethiopia is based on regional ethnic federal structure with constitutional right to secede for any ethnic group. One ethnic group dominates others based on a plan (CIA 1985). According to the constitution, power is shared based on the population of ethnic groups, but the TPLF (Tigray Peoples Liberation Front) allegedly controls the power despite its representation of 6% ethnic minority in the north, Tigray (The Economist, 13 October 2016).

The Ethiopian constitution reserves right for fragmentation of ethnic regions where ownership of big projects will be determined by their mere location. The 1995 Ethiopian constitution is quoted here to show the legalization of potential brake down of the country on ethnic basis, Article 39(1–4). Article 39(4c and d) states that a law will be enacted for distribution of property at time of secession of an ethnic region which the power at the time will make the rule.

"Article 39
Rights of Nations, Nationalities, and Peoples
1. Every nation, nationality and people in Ethiopia has an unconditional right to self determination, including the right to secession.
2. Every nation, nationality and people in Ethiopia has the right to speak, to write and to develop its language; to express and to promote its culture; and to preserve its history.
3. Every nation, nationality and people in Ethiopia has the right to a full measure of self-government which includes the right to establish institutions of government in the territory that it inhabits and to equitable representation of regional and national governments.
4. The exercise of self-determination, including secession of every nation nationality and people in Ethiopia is governed by the following procedure: (a) When a demand for secession has been approved by a two-thirds majority of the member of the members of legislative council of any nation, nationality or people; (b) When the Federal Government has organized a referendum which must take place within three years from the time it received the concerned Council's decision for secession; (c) When the demand for secession is supported by a majority vote in the referendum; (d) When the Federal Government will have transferred to the people or to their Council its powers and; (e) When the division assets is effected on the basis of law enacted for that purpose.

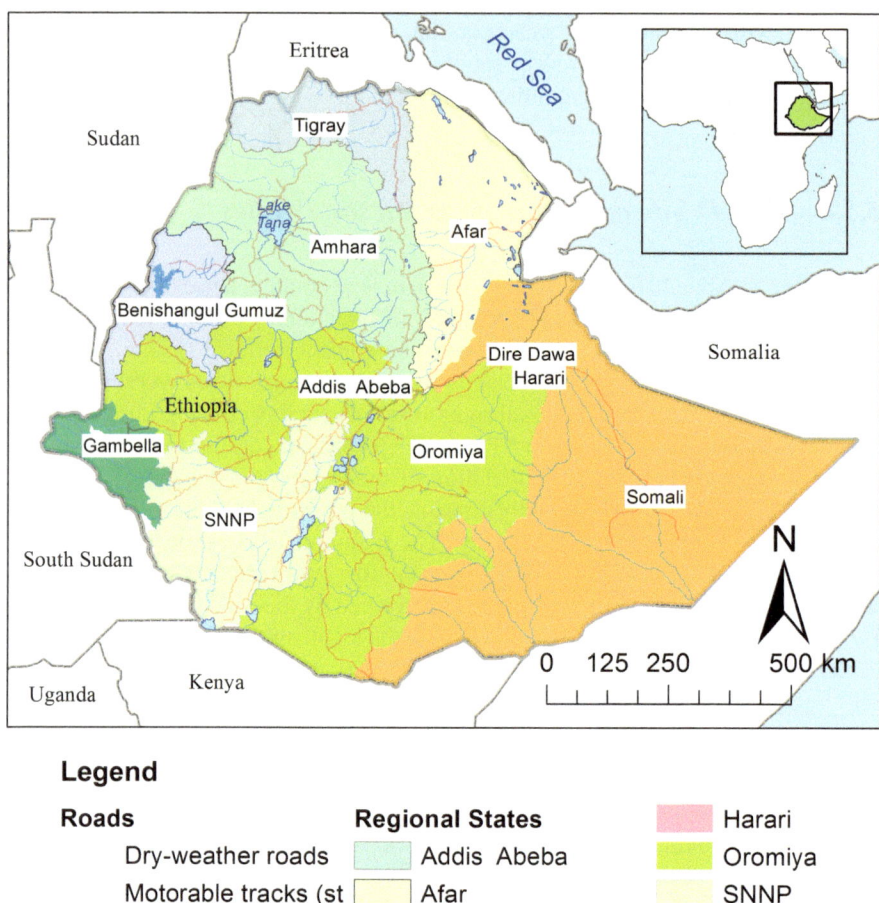

Fig. 5.6 Administrative regions/states of Ethiopia based on dominant ethnic group

5. A nation, nationality or people for the purpose of this Constitution, is a group of people who have or share a large measure of a common culture, or similar customs, mutual intelligibility of language, belief in a common or related identities, and who predominantly inhabit an identifiable, contiguous territory."

The current site of the dam is in a small minority region, the Benshangul-Gumuz, outside the Amhara and Oromo regions of major ethnic groups. The main source

(a)   (b)

Fig. 5.7 **a** Water harvesting pool digging in the Blue Nile basin upstream of GERD; **b** Water harvesting pool with plastic liner in the Nile basin

of the Blue Nile is the Amhara region followed by the Oromia region (Fig. 5.6). In a system which operates on ethnic identity, the site selection may have major implications on who will own and benefit from the construction to power generation and water based economic gains of the dam.

Food shortage and lack of drinking water is common in many areas in the Blue Nile basin. The eastern Amhara region is drought prone and water scarce where communal water harvesting ponds are necessity. Shallow wells, springs and streams are source of water supply (Deneke 2014). There was a national program of local water harvesting through catching runoff in ponds to relief food and water shortage in the Nile basin in Ethiopia that was aggressively expanded (Fig. 5.7a, b). However, how much the program is currently supported is not clear. Benefits and associated problems of water harvesting mainly in the form of ponds has been reported (Tesfay 2011; Mume 2014)

The Blue Nile and other Ethiopian rivers are trans-ethnic within Ethiopia with potential internal water conflict as much as with downstream users outside Ethiopian border. The dam has a potential to be a cause of conflict within Ethiopia for land, water, power and reservoir associated developments and upstream water rights as long as the current ethnic based constitution is enacted. Upstream water rights are in conflict with the dam objectives and no recognition of the issue and consultation of upstream people is reported. Growing upstream dependence on irrigation and water abstraction could soon surface as water right issue.

5.5.5 Summary

The GERD dam site selection presents complex and intertwined technical, environmental, economic and socio-political challenges. The site maximized the flow volume of the Nile River basin to be used for hydroelectric power generation and reduced the prospective transmission cost in future power trade with Sudan and Egypt. The geology of the site has not only enabled the strong foundation to build the largest gravity dam in the region but also provided construction material for the 5 km long rockfill dam. The dam site arguably brought up domestic and regional socio-political and environmental issues. Ethiopian ethnic political structure are intertwined with potential internal conflict for the ownership of the dam and its benefits has been argued as a major domestic socio-political factor. The site selection importance will be recognized as time goes. The closeness of the dam to Ethio-Sudan boarder has been argued by Ethiopia as expression of co-operation but also considered by as disadvantaged in terms of security due to its location and proximity to international borders. With downstream riparian water right and upstream water right and continuous increase in water demand and withdrawal at both ends, the location of GERD could be a reason bring all parties together for dialogue towards co-operation and regional development.

References

Abtew W, Melesse AM, Dessalegne T (2009) Spatial, inter and intra-annual variability of the Upper Blue Nile Basin rainfall. Hydrol Process 23:3075–3082

Abdelhady D et al (2015) The Nile and the Grand Ethiopian Renaissance Dam: is there a meeting between nationalism and hydrosolidarity? J Contemp Water Res Educ 1555(1):73–82

Berhanu B, Sileshi Y, Melesse AM (2014) Chapter 6: Surface water and groundwater resources of Ethiopia: potentials and challenges of water resources development. In: Melesse AM, Abtew W, Setegn (eds) Nile River Basin ecohydrological challenges, climate change and hydropolitics. Springer, New York

CIA (1985) Special issue: insurgencies in Sub-Saharan Africa. African Review, pp 13–14 (Declassified in part-sanitized copy approved for release 2012/07/06). https://www.cia.gov/news-infor mation/press-releases-statements/2017-press-releases-statements/cia-posts-more-than-12-millio n-pages-of-crest-records-online.html

Deneke TT (2014) Chapter 24 Processes of institutional change and factors influencing collective action in local water resources governance in the Blue Nile Basin of Ethiopia. In: Melesse AM, Abtew W, Setegn SG (eds) Nile River Basin ecohydrological challenges, climate and hydropolitics. Springer, New York

Gebreluel G (2014) Ethiopia's Grand Renaissance Dam: ending Africa's oldest geopolitical rivalry? Washington Q 37(2):25–37

GSE (1996) Geological map of Ethiopia, Go1:2,000,000 scale, 2nd edn. http://www.gse.gov.et

International Rivers (2012) Field visit report GERD Project. August 2012, Ethiopia. https://www.i nternationalrivers.org/sites/default/files/.../grandren_ethiopia_2013.pd

IPoE (2013) International panel of experts on Grand Ethiopian Renaissance Dam project. Final report. Addis Ababa, Ethiopia

Legesse E (2014) Yemelese trufatoch balebet alba ketema (Amharic). Netsanet Publishing Agency, U.S.A

Mume J (2014) Impact of rain-water-harvesting and socio economic factors on household food security and income in moisture stress areas of eastern Hararghe, Ethiopia. Int J Novel Res Market Manage Econ 1(1):10–23

Pearce F (2012) The land grabbers: the new fight over who owns the earth. Beacon Press Boston, MA investment

Saeed SY (2018) Ethiopia's Renaissance Dam and its impact on Sudanese water security. Sudan Tribune, 7 Jan 2018

Tadesse S, Milesi J-P, Deschamps Y (2003) Geology and mineral potential of Ethiopia: a note on geology and mineral map of Ethiopia. J Afr Earth Sci 36:273–313

Tesfay G (2011) On-farm water harvesting for rainfed agriculture development and food security in Tigray, Northern Ethiopia. DCG Report No 61

Van Der Schalie (1974) Aswan Dam revisited. Environ Sci Policy Sustain Dev 16:18–20. https://doi.org/10.1080/00139157.1974.9928525

Veilleux JC (2013) The human security dimensions of dam development: the Grand Ethiopian renaissance Dam. Global Dialogue 15(2): Summer/Autumn 2013—Water cooperation or conflict?

Veilleux JC (2014) Is dam development a mechanism for human security? Scale and perception of the Grand Ethiopian Renaissance Dam on the Blue Nile River in Ethiopia and the Xayaburi Dam on the Mekong River in Laos. Ph.D. dissertation, Oregon State University, Corvallis, Oregon

Chapter 6
Grand Ethiopian Renaissance Dam Analysis

Abstract The Grand Ethiopian Renaissance Dam is the first major dam in the Blue Nile (Abay) River of Ethiopia. GERD is a combination of 175 m high roller compacted concrete gravity dam and a 50 m high concrete faced rock fill saddle dam under construction by Ethiopia. The gravity dam is built across the natural course of the Blue Nile River and the saddle dam provides the design storage and water level due to the relatively low relief of the dam site. The dam is being built at the most downstream site of one of the four potential dam sites proposed by a 1964 feasibility study of the Blue Nile basin development conducted by United States Bureau of Reclamation. The dam has been under construction since 2011 with 70% completed at the beginning of 2018. The installed power generation capacity of 6,000 MW is expected to be generated by 16 Francis Turbines each with 375 MW capacity located at the foot of the main dam. The design flow rate of 4305 m^3 s^{-1} is about 3 times the average flow. At the average flow rate, the expected average annual energy production is 15,700 GWH. The suitability of the dam site, dam design, major components and operations are discussed.

Keywords Grand Ethiopian Renaissance Dam · Ethiopia · Blue Nile river
Dam design · Dam operation · Ethiopia · Sudan · Egypt

6.1 Introduction

The Grand Ethiopian Renaissance Dam (GERD) is the first major dam on the Blue Nile River (Abay) in Ethiopia. There are two operational dams (Roseires and Sennar) on the Blue Nile River in the Sudan and several other dams on the Nile River and tributaries in the Sudan and Egypt (Fig. 6.1). The volume of flow and topography along the Blue Nile River in Ethiopia presents a tremendous hydroelectric power potential. The Blue Nile River originates from the northeast part of the watershed and travels clockwise about 1000 km from its source of Gilgel Abay upstream of Lake Tana to the Sudan border. The river falls 1295 m to the Sudan border (490 m asl), beginning with the spectacular Tis'Esat Falls about 30 km downstream of Lake Tana outlet (1785 m asl).

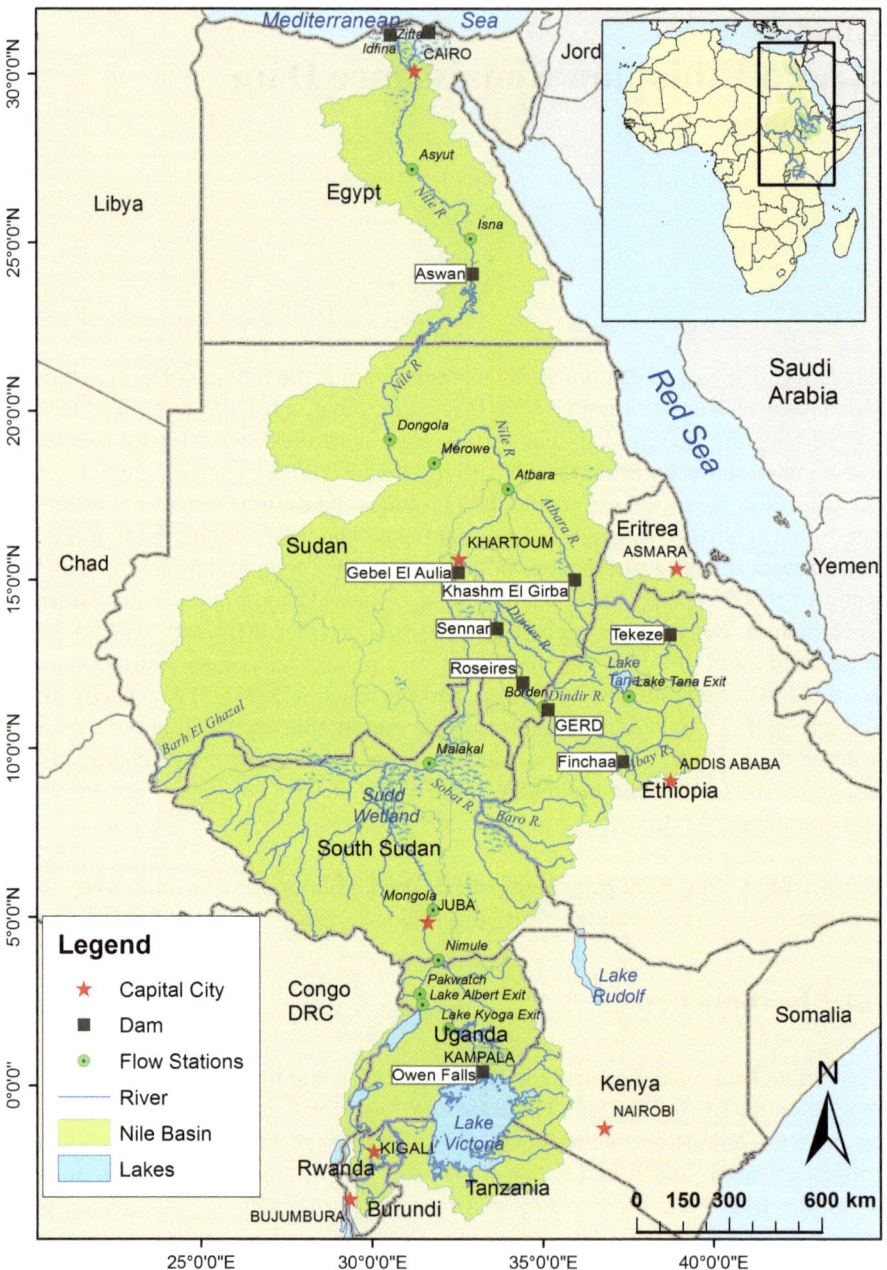

Fig. 6.1   The Nile River Basin and location of major Dams

There is a more than 10 fold increase in discharge between Lake Tana and the Sudan border. Average rainfall in the Blue Nile basin is 1423 mm with standard deviation of 125 mm (Abtew et al. 2009). Rainfall drastically decrease at the border and further north turning into desert environment. The construction of the GERD on the Blue Nile River is a reflection of severe energy shortage for economic development to address the needs of an exploding population growth.

Despite the large drainage contributing to the Nile Basin, Ethiopia has only two operational hydropower dams in its Nile sub-basins; one in the Tekeze (Atbara) basin and the other Finchaa hydropower dam on a tributary of Abay (Blue Nile) River. GERD is the first dam on the main Blue Nile river of Ethiopia. There is no major dam on the Ethiopian Baro-Akobo tributary of the Nile River. Ethiopia has been contemplating construction of dams on the main Blue Nile River for a very long time. During the construction of the High Aswan Dam in Egypt with the support of the Soviet Union, the United States Bureau of Reclamation (USBR) studied potential hydropower dam sites on the Blue Nile in Ethiopia from 1956 to 1964. In 1964, it offered four proposed dam sites, Karadobi, Mabil, Mendia and Border. Figure 6.2 depicts proposed dam sites and the topography of the Blue Nile basin with relatively low relief at the Ethio-Sudan border. The current Ethiopian Government selected the Boarder site and began construction of the GERD in April 2011. The sole purpose of the dam was announced to be hydroelectric power for consumption in Ethiopia and export to Egypt and Sudan. GERD is being constructed near the Boarder site almost half a century after the initial feasibility study. The GERD reservoir is expected to reach as far back as the proposed Mendia site at Full supply level.

The High Aswan Dam (HAD) in Egypt is the largest dam in the Nile River Basin with a reservoir capacity of 132 bcm (billion cubic meter). Upon completion in 1970, the HAD was intended for sustainable irrigation development, hydropower and navigation improvement. The dam has been credited to save Egypt from extreme floods and droughts that devastated the upstream riparian nations (Abu-Zeid and El-Shibini 2010; Strzepek et al. 2008).

The Ethiopian government has argued that the closeness of the dam to the boarder is re-assurance to downstream countries that the purpose of the dam is non-consumptive hydroelectric power generation without reducing the annual volume of water flowing to Sudan and Egypt due to the construction of the GERD. GERD site considerations are presented in Chap. 5. This chapter presents analyses of the design of the dam and main reservoir features with respect to site selection, operational considerations and power generation.

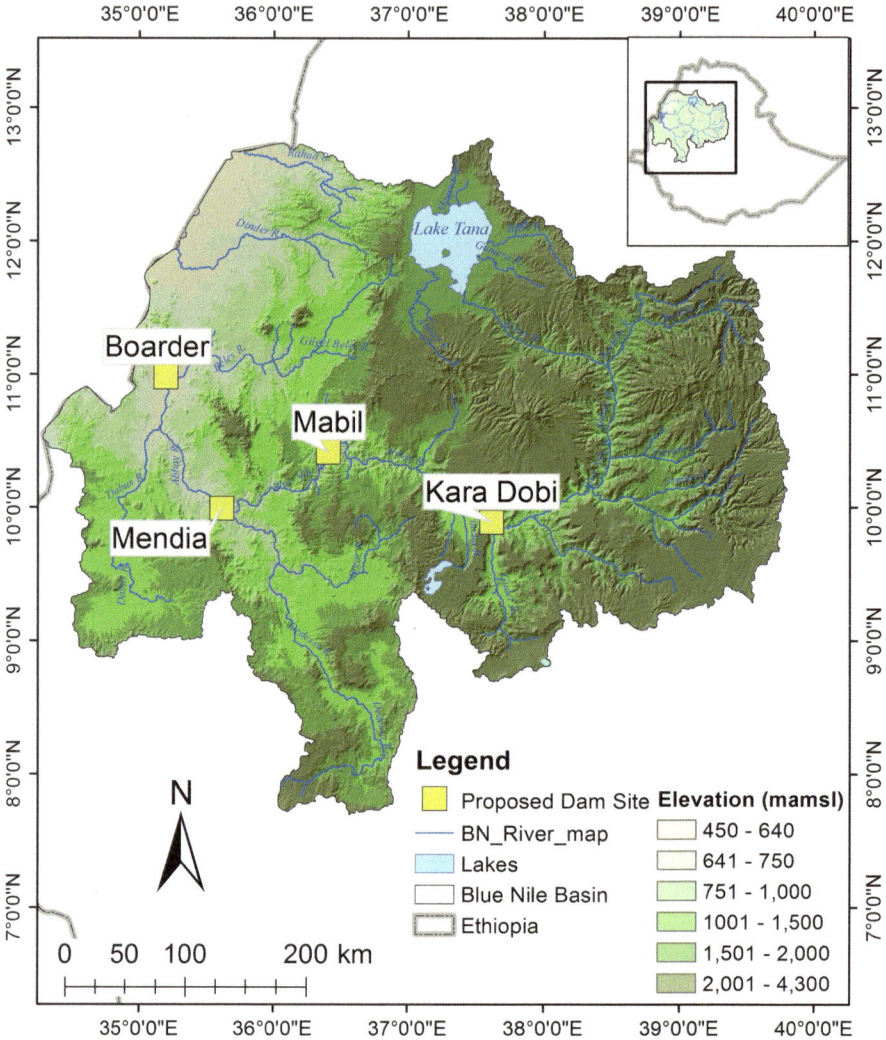

Fig. 6.2 Blue Nile River basin topography and location map of the four hydroelectric dam sites proposed in 1964; Kara Dobi, Mabil, Mandia, Border (GERD)

6.2 Dams on the Nile

Water control on the Nile goes as far back as 2650 BC when Sadd el-Kafara, a masonry dam, that was built to control flooding but failed due to high flows, leaving its mark as the oldest dam of its size (Mays 2010). The earliest major dam is the Low Aswan Dam, a masonry gravity dam in Aswan, Egypt, built between 1898 and 1902, 1000 km south southeast of Cairo (Inman and Jenkins 1985). The Low Aswan dam was 38 m high after being raised twice in 1907 and 1929, and raised the water level of the Nile River for a stretch of 350 km upstream (Inman and Jenkins 1985). The capacity of the Low Aswan Dam was 980 million m^3 (The New York Times, 27 July 1913). Dams and water control structures on the Nile is presented in Table 6.1. The major purposes of irrigation and hydropower are provided with storage capacity where information is available.

There are over 25 dams and major water control infrastructures on the Nile River. The Sennar and Roseires dams in Sudan are on the Blue Nile. The High Aswan Dam (HAD) in Egypt is the largest dam in the Nile River Basin with a reservoir capacity of 132 bcm. The HAD has been perceived as a political symbol and achievement (Van Der Schalie 1974). The GERD is the first and only dam being built on the Blue Nile river of Ethiopia. But, water rights issues have generated downstream resistance to the GERD. The major factor for downstream concern is potential flow reduction and lack of control of the dam operation. Looking at the regional population and economic growth, the number of existing dams on the Nile will likely increase in the coming decades to meet the growing energy and water demand in the basin and GERD is just one of them.

6.3 The GERD

The GERD is comprised of a main gravity dam, a gated main spillway, and a rock-fill dam located about 20 km from Ethio-Sudan boarder. The purpose of the GERD is stated as hydroelectric power generation to boost Ethiopia's domestic power supply and power export to neighbouring countries. The dam site is located about 500 km northwest of the capital Addis Ababa. Since the dam site is further downstream of the Blue Nile gorge, the topography of the dam site requires combination of gravity dam and rock-fill saddle dam to generate the designed storage reservoir. The dam will raise the water level from 500 m asl at the natural river surface to 640 m asl at its full supply level. The reservoir stretches about 246 km upstream from the dam site. The natural river bed at the dam site is 500 m above mean sea level. Table 6.2 depicts some of the site characteristics and drainage or watershed area.

6.4 Dam Design Characteristics

Dam construction started in 2011 with about 70% completed by the beginning of 2018 according to Ethiopian Media and 60% as reported by Reuters (Reuters 1 January 2018). The main dam is roller-compacted concrete (RCC) gravity dam. The difference between RCC and conventional concrete dam is that RCC has time advantage not requiring as much cooling and forming time. It has also advantage on cost. By 2008, there were 373 RCC dams around the world with height higher than

Table 6.1 Dams and water control structures on the Nile River and tributaries (Tesemma 2009; Conniff et al. 2012)

Dam/Water control structure	Built	Country	Purpose, irrigation (I); hydropower (HP in MW)	River
Delta Barrage	1833–1862	Egypt		Nile
Aswan Low Dam	1898–1902	Egypt	I (5.1 bcm)	Nile
Assuit Barrage	1903	Egypt	I	Nile
Nag-Hamady	1930	Egypt	I	
Aswan High Dam	1960–1970	Egypt	I (132 bcm) and HP (2100)	Nile
Al-Sallam Canal	1997	Egypt	Transfer water to Sinai	Nile
Finchaa	1973	Ethiopia	HP (84)	Blue Nile tributary
CharChara	2000	Ethiopia	HP(73)	
Koga	2008	Ethiopia	I; HP(80)	Blue Nile tributary
Tekeze	2009	Ethiopia	HP (300)	Atbara
Tana-Beles	2010	Ethiopia	I (Lake Tana) HP (460)	Blue Nile
GERD	Under construction	Ethiopia	HP (6000)	Blue Nile
Megech		Ethiopia	I	Inflow to Lake Tana
Chemoga Yeda	2013	Ethiopia	278	Blue Nile tributary
Gumera		Ethiopia	I	Inflow to Lake Tana
Ribb	Under construction	Ethiopia	I	Inflow to Lake Tana
Didessa	Under construction	Ethiopia	I	Blue Nile tributary
EwasoNgiro	2012	Kenya	HP (180)	
Sennar	1925	Sudan	Irrigation (0.8 bcm)	Blue Nile
Jebel Aulia	1937	Sudan	summer irrigation for Egypt	Nile
Kash el Girba	1964	Sudan	I (1.3 bcm) HP (10)	Atbara
Roseires	1966	Sudan	I (7.4 bcm) HP (1800)	Blue Nile
Merowe	2009	Sudan	I 12.5 bcm HP (1250)	Nile
Dal	Planned	Sudan	HP (400)	Nile
Shereyk	Planned	Sudan	I HP (350)	Nile
Kajbar	In progress	Sudan	I HP (30)	Nile

(continued)

Table 6.1 (continued)

Dam/Water control structure	Built	Country	Purpose, irrigation (I); hydropower (HP in MW)	River
Rumela Dam	2011–	Sudan	I HP 135	Atbara
Burdana Dam	2011–	Sudan	I HP 120	Setit
Jonglei Canal	1978 not completed	Sudan	Drain the Sudd wetlands	White Nile
Rusumo I & II	2012	Rwanda	HP (60)	Kagera
Nyabarongo	2012	Rwanda	Hp (27)	Kagera
Owens Fall (Nalubaale)	1954	Uganda	HP (180)	Nile (Lake Victoria)
Kiira	2000	Uganda	I HP 2000	Nile tributary
Isimba	2015	Uganda	HP(87)	White Nile

Table 6.2 Dam site and basin (IPoE 2013)

Basin and dam site	
Drainage area	172,250 km^2
Site upstream elevation	640 m asl
Site downstream elevation	500 m asl
Length of reservoir site (full)	246 km
Estimated slope	61 cm km^{-1}

15 m. Forty three of these dams are in the United States with the tallest being the Upper Stillwater Dam (90 m) with RCC volume of 124,600 m^3 (Abdo 2008).

GERD is composed of two types of dams, roller compacted concrete gravity main dam and a rock filled saddle dam. The spillway is also a separate structure from both the main dam and the saddle dam. Hence, GERD comprises a Gravity dam on the main river channel, Spillway on the right, and an embankment dam. The gravity dam is constructed on the natural river course between two hills (Figs. 6.3 and 6.4a).

A 5 km long saddle dam is necessary to maintain the required water surface elevation and depth at the relatively flat dam site (Fig. 6.4b). The main dam across the river has three sections. The left power house, the central block and the right power house. There are two spillways in the central block, one gated and one ungated. There are four bottom outlets. Table 6.3 depicts the design features. Figure 6.5 depicts the profile and major features of the GERD outlets and operational water levels based on information compiled from multiple sources.

Fig. 6.3 GERD main dam as displayed on Goggle Earth showing the status of the Rockfill saddle dam and the Roller Compacted Concrete (RCC) gravity dam in 2016

6.4.1 Main Gravity Dam

The main dam is roller-compacted concrete (RCC) gravity dam constructed on the natural river course. The dam is constructed on a straight 1800 m axis between two hills maintaining its stability against design loads from the geometric shape, and the mass and strength of the concrete. Like most gravity dams, the GERD main dam consists of a non-overflow section and a central overflow spillway section and houses two power houses on the left and right foot of the dam (Figs. 6.4a and 6.5). The dam also contains sixteen penstocks and four diversion (bottom) outlets (Fig. 6.5).

Fig. 6.4 **a** GERD RCC gravity dam showing the main dam and the left side power house under construction; **b** concrete faced rockfill saddle dam (2018)

6.4.2 The Saddle Dam

The 50 m high and 5 km long concrete face rockfill saddle dam was necessary to maintain the required water surface elevation and depth at the relatively flat dam site (Fig. 6.4b). The saddle dam raises the natural land feature from 600 m asl to 646 m asl raising the reservoir water level to the design level. The saddle dam is a rock filled dam with concrete upstream face (Fig. 6.6a, b). As a result of site topography, the saddle dam is needed to be 5 km long covering the gap between two hills. An emergency gated 300 m wide spillway is located between the main dam and the saddle dam (Fig. 6.7). The spillway, at crest elevation of 624.9 m, is to be used for

Table 6.3 GERD main dam design features

Main dam design		
Dam height (max)	155 m (Salini)	
Elevation	640 m asl	
Length	1800 m (Salini)	
Design flow	4305 m^3 s^{-1} (Salini)	
Maximum net head	133 m (IPoE 2013)	
Dam Volume	10.2 m^3 x 10^6 (Salini)	
Main spillway (gated) 6 sluices (14 × 5.5 m)	15,000 m^3 s^{-1} (Salini)	Sill elevation 624.9 m asl; 300 m long (IPoE 2013)
Emergency spillway (ungated) sill elevation 640 m asl; 205 m long	for 10,000-year flow (IPoE 2013)	on the crest of central block
4 gated outlets (8 m D and 210 m L)	for diversion during construction and dewatering later	
Right shoulder	power house (10 Francis turbines)	375 MW each (IPoE 2013)
Left shoulder	power house (6 Francis turbines)	375 MW each (IPoE 2013)
Average energy production per year	15,700 GWH	(Salini)
Total installed capacity	6000 MW	(Salini)
Plant factor (IPoE 2013)	0.31 (ratio of actual to potential output)	

extreme flood conditions releasing through a gully into the river downstream of the dam. Table 6.4 depicts features of the saddle dam.

6.5 Reservoir

The reservoir is a large water body half the area of the upstream Lake Tana in Ethiopia, when full. The reservoir area is subject to fluctuation between 522 and 1736 km^2, between the Full Supply Level (640 m asl) and Minimum Operating Level (590 m asl). Water surface elevation, area and volume are important parameters in analysis of reservoir filling and operation. In this chapter, these parameters were derived from 30 m SRTM digital elevation model (Jenson and Domingue 1988). The area to be flooded is depicted in Fig. 6.7 with the main dam and saddle dam marked. Table 6.5 and Fig. 6.8 show water surface elevation, area and volume of reservoir. The reservoir has a volume of 73 bcm. The operating range is 50 m and the dead storage is 14.8 bcm (IPoE 2013).

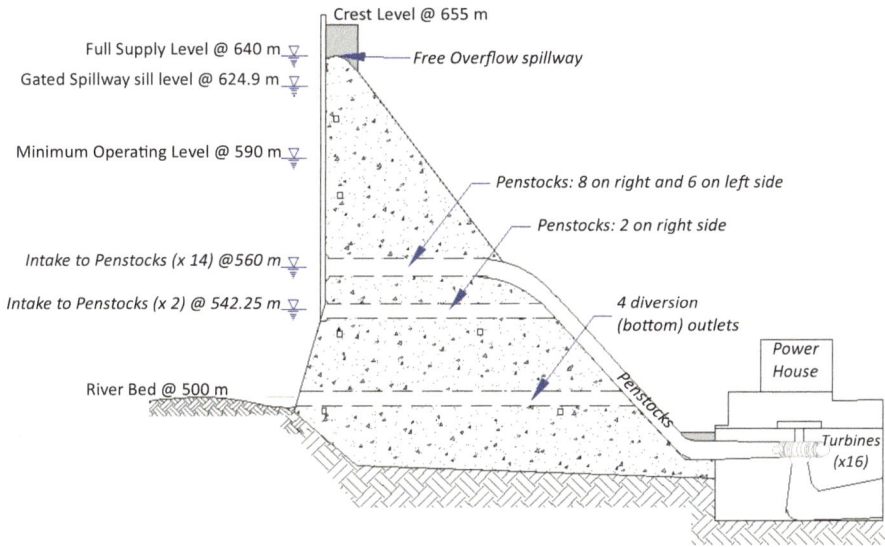

Fig. 6.5 GERD main dam outlets and operational water levels

6.6 Flow Characteristics and Flow Data

Flow characteristics and sediment load are important parameters in dam site selection and design. Blue Nile River flow seasonal distribution varies with rainfall distribution and flood travel time. Average annual flow of the Blue Nile (Abay) at Ethio-Sudan boarder is approximately 50 bcm. Flow temporal variation patterns rainfall variation except river flow lags with distance from the runoff generating sub-basins. Flood flow at Ethio-Sudan border and Sennar dam downstream is from July to October while it is August to November at Aswan dam. Ethio-Sudan border and Sennar flows are from the Blue Nile while Aswan receives additional flows from the White Nile, Sobat and Atbara. July through October flow at the dam site is 80% of the annual flow (Abtew et al. 2009). Blue Nile river characteristics at the dam site are depicted in Table 6.6. Mean monthly flow at Sennar dam and border comparing multiple data sources are shown in Fig. 6.9.

Published annual flow data estimates for the Blue Nile River at the border with Sudan and at Roseires dam downstream in Sudan vary with reporting. Mulat et al. (2014) report average Blue Nile annual flow at GERD site as 47.1 bcm (1967–72). FAO reports 48.7 bcm average annual flow at El Diem (1912–1997), close to border (FAO accessed 24 March 2018). Whittington et al. (2015) reported an average annual flow at GERD site of 48 bcm. Average annual flow at Roseires dam of about 41.7 is derived from graphic report by Keith et al. (2014) for the period 1971–1991. Another estimate of average annual flow at Roseires from graphic report (1912–1991) by Tesemma (2009) is estimated as 50 bcm. Five years flow data at the Sudan border (GERD site) from Ethiopian Meteorological Agency (1999–2003) has mean annual

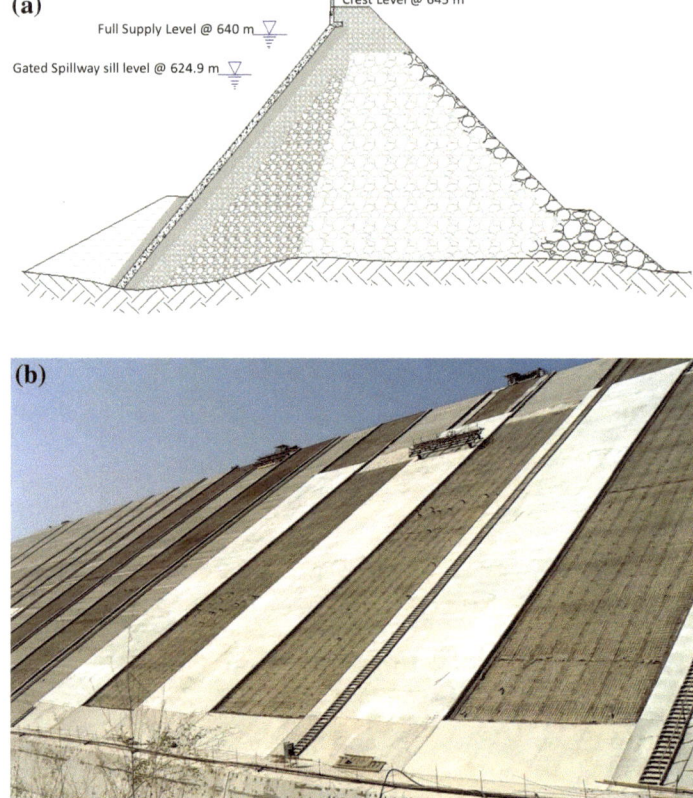

Fig. 6.6 **a** GERD saddle dam profile; **b** concrete faced rockfill saddle dam construction progress of the upstream concrete face (2018)

flow of 53.7 bcm. The combined annual average flow estimate at GERD site from the above two sources (1967–1972; 1999–2003) is 50 bcm.

6.7 Does the GERD Has 6000 MW Capacity?

The total installed power generation capacity of GERD is 6000 MW as reported in the design. Available head and flow rate mainly control the amount of power generated by hydropower plants. Head is the depth of water in the reservoir above the intake to the turbines. If incoming flow rate into the reservoir is low and outflow to turbines is high, the system loses head as reservoir water level goes down. The design flow rate of 4305 m^3 s^{-1} is 2.78 times the average stream flow. Accordingly, the expected average annual energy production 15,700 GWH.

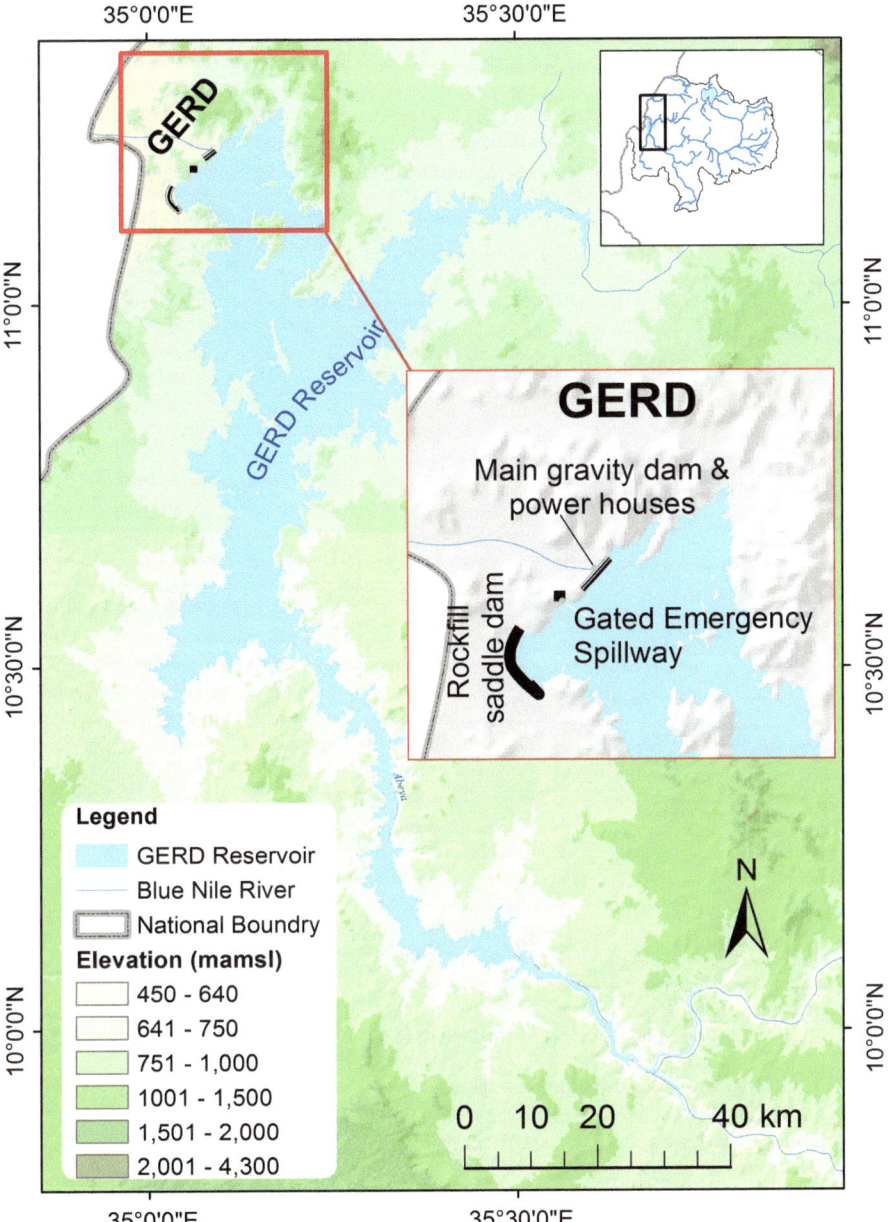

Fig. 6.7 GERD reservoir surface area with location of dam and saddle dam

Table 6.4 GERD saddle dam features

Saddle dam (Salini)	
Dam height	50 m
Length	5 km
Crown elevation	645 m asl
Emergency spillway (ungated)	Sill elevation 642 m asl
Dam body volume	16.5 million m^3

Table 6.5 Reservoir water surface elevation, area and volume relationship as derived from 30 m DEM

Elevation (m asl)	Area (km^2)	Volume (MCM)
645	1883	73,745
640	1736	64,773
625	1304	42,208
590	522	11,998
565	204	3442
560	164	2548
542	61	675

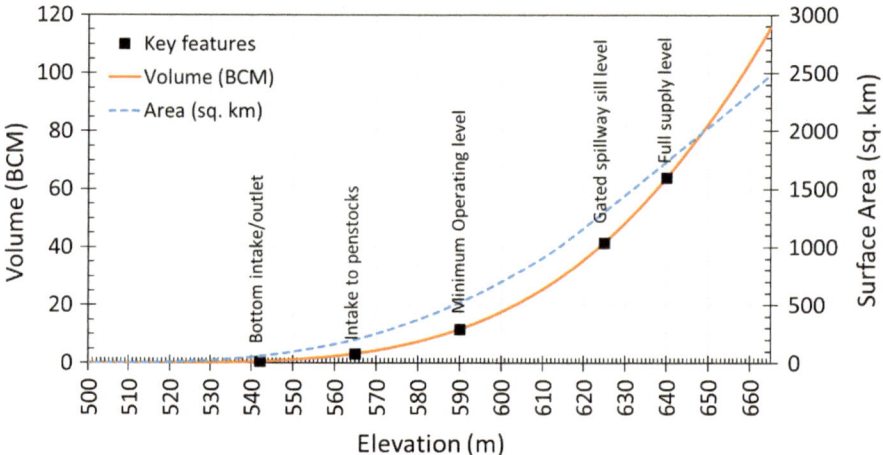

Fig. 6.8 GERD reservoir elevation-area-volume curve

Table 6.6 Blue Nile River design flows at GERD dam site (IPoE 2013)

River characteristics	
Mean flow	1547 m^3 s^{-1}
10-year flow	14,700 m^3 s^{-1}
10,000-year flow at 641.8 m asl	26,860 m^3 s^{-1}
Probable Maximum Flood at 642.9 m asl	38,750 m^3 s^{-1}
Mean annual sediment yield	207 million m^3

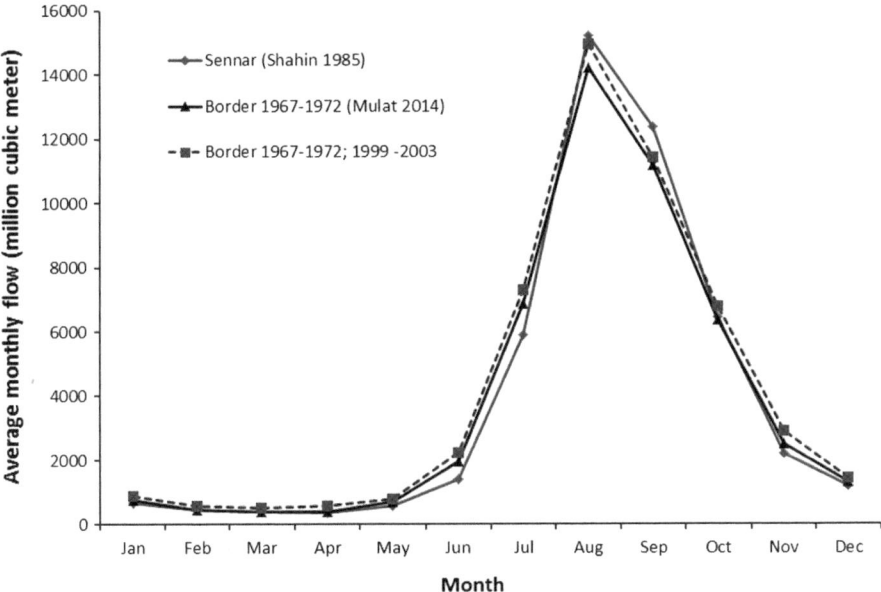

Fig. 6.9 Mean monthly flow at border and Sennar dam

Table 6.7 Four dams in series on the Blue Nile (data derived from Block et al. 2007)

Project	Dam height (m)	Reservoir capacity (billion m^3)	Design head (m)	Flow at design head (m^3 s^{-1})	Installed power at design head (MW)
Karadobi	252	32.5	181.4	948	1350
Mabil	171	13.6	113.6	1346	1200
Medaia	164	15.9	117.4	1758	1620
Border	84.5	11.1	75	2378	1400
Total		73.1			5570

Block et al. (2007) performed hydropower and irrigation modelling at the four proposed dam sites by USBR; Karadobi, Mabil, Mandia and Border (Fig. 6.2). They estimated installed power at design head of a combined total 5570 MW for the four dams in series (Table 6.7).

The power equation is applied to varying head and discharge rate to estimate turbine generated power (Eq. 6.1). Ranges of flow rate (Q) between mean flow 1547 m^3 s^{-1} and design flow 4305 m^3 s^{-1} are applied to demonstrate the effect of reservoir water level decline on power generation for a given flow rate (International Rivers 2013).

$$P = \frac{\eta \rho Q g h}{1000000} \tag{6.1}$$

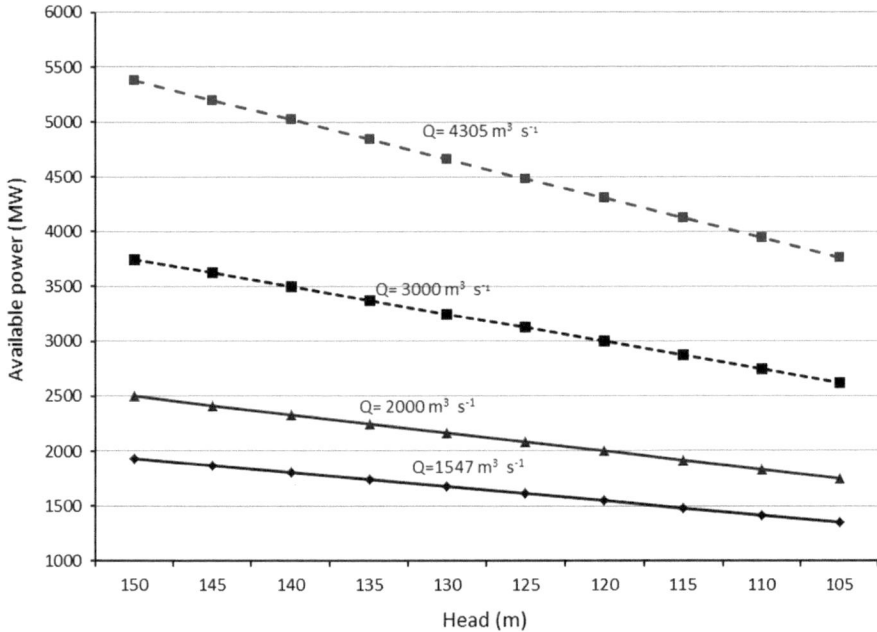

Fig. 6.10 Hydropower capacity with available head and flow rate

where P is power in MW, n is turbine efficiency (0.85), p is density of water (1000 kg m^{-3}), Q flow rate (m^3 s^{-1}), g is gravitational acceleration (9.81 m s^{-1}) and h is head in m. Figure 6.10 depicts available power from GERD operation for average flow rate of 1547 m^3 s^{-1} (IPoE 2013) and design flow of 4305 m^3 s^{-1} from Salini-Impregilo (Accessed 15 October 2016), the company in charge of design and construction of GERD; and for varying head or water level. In between the two discharge rates, 2000 and 3000 m^3 s^{-1} discharge rates were also applied.

6.8 Dam Design Review by International Panel of Experts (IPoE)

An International Panel of Experts (IPoE) was formed and studied the design of the GERD between 2012 and 2013, visited the dam site, and produced final report (IPoE 2013). The panel constituted two experts each from Egypt, Ethiopia and Sudan and four international experts. The two widely discussed topics regarding the GERD were (1) detail of structural design of the dam, (2) the implications of filling of the dam. Generally, the review comments are guarded with implicit acknowledgement of the fait accompli status of the dam design and construction stage and provided recommendations. In several cases it is stated that complete design documents and test data were not available for review. According to the review, upstream and down-stream flow and meteorological data were not available. Increasing the gated spillway and dropping the low discharge capacity (93 m³ s⁻¹ during the 10,000-year flood) side channel spillway on the saddle dam, was recommended in return increasing the capacity of the gated spillway. The reduction of the design PMF (Probable Maximum Flood) from 38,750 m³ s⁻¹ to 30,200 m³ s⁻¹ was not supported by the IPoE. Gated spillway erosion potential concern was stated. It recommended reservoir water quality monitoring with respect to dissolved oxygen, organic matter and sediments. The IPoE recommended two studies on the Eastern Nile System, (a) water resource system study and hydropower modelling and (b) transboundary environment and socioeconomic impact studies through appropriate arrangement as agreed by the three countries, by employing renowned international consultants through international bidding process.

6.9   Summary

GERD is a combination of 1.8 km long roller-compacted concrete gravity dam, a 300 m long spillway and a 5 km embankment saddle dam required to maintain the design water level and storage given the relatively flat topography of the dam site. Stage-area-volume curves are developed to show area under water for given water level (stage) showing the saddle dam is a necessity to provide the design storage and water level for hydropower generation at the site. The location close to an international border increases security risk of the dam. The design of spillways and overall construction will be put to the test during filling and operation. The design parameters of the dam and sketches are provided which are later used for filling period evaluation and operation of the dam. Power generation capacity is evaluated with varying flow rate and heads.

References

Abdo FY (2008) Roller-compacted-concrete dams: design and construction. Hydro Review, HCI Publications, pp 1–5
Abtew W, Melesse AM, Desalegn T (2009) Spatial, inter and intra-annual variability of the Upper Blue Nile Basin rainfall. Hydrol Process 23:3075–3082
Abu-Zeid MA, El-Shibini FZ (2010) Egypt's high Aswan dam. Int J Water Resour Dev 13:209–218. https://doi.org/10.1080/07900629749836
Block PJ, Strzepek K, Rajagopalan B (2007) Integrated management of the Blue Nile basin in Ethiopia. IFPRI Discussion Paper 00700. Colorado University, Boulder, Co
Conniff K, Molden D, Pedan D, Awulachew SB (2012) Nile water and agriculture past, resent, future. In Awulachew SB, Smakhtin V, Molden B, Pedan D. The Nile River basin. Routledge Taylor and Francis Group, New York
FAO Hydrologic regime in the Nile basin. http://www.fao.org/docrep/015/an530e/an530e.pdf. Accessed 24 Mar 2018

Inman DL, Jenkins SA (1985) The Nile littoral cell and man's impact on the coastal zone of the southeastern Mediterranean. In: Coastal Engineering 1984, pp 1600–1617

International Rivers (2013) Ethiopia's biggest dam oversized, experts say. Interview with Professor Asfaw Beyene. Sept 5, 2013. International Rivers

IPoE (2013) International panel of experts on Grand Ethiopian Renaissance Dam project. Final Report, Addis Ababa, Ethiopia

Jenson SK, Domingue JO (1988) Extracting topographic structure from digital elevation data for geographic information system analysis. Photogr Eng Remote Sensing 54:1593–1600

Keith B, Epp K, Houghton M, Lee J, Mayville R (2014) Water as a conflict driver: estimating the effects of climate change and hydroelectric dam diversion on the Nile River stream flow during the 21st century. Center for Nation Reconstruction and Capacity Development. United States Military Academy, West Point, New York

Mays LW (2010) Water technologies in ancient Egypt. In: Mays LW. Ancient water technologies. Springer, New York

Mulat AG, Moges SA, Ibrahim Y (2014) Chapter 27 impact and benefit study of Grand Ethiopian Renaissance Dam (GERD) during impounding and operation phases on downstream structures in the Eastern Nile. In: Melesse AM, Abtew W, Setegn S (eds) Nile River Basin ecohydrological challenges, climate change and hydropolitics. Springer, New York

Salini-Impregilo. Grand Ethiopian Renaissance Dam Project. http://www.salini-impregilo.com/en/projects/in-progress/dams-hydroelectric-plants-hydraulic-works/grand-ethiopian-renaissance-dam-project.html. Accessed 2 Jan 2017

Strzepek KM, Yohe GW, Tol RSJ, Rosegrant MW (2008) The value of the high Aswan Dam to the Egyptian economy. Ecol Econ 66:117–126. https://doi.org/10.1016/j.ecolecon.2007.08.019

Tesemma ZK (2009) Long term hydrologic trends in the Nile basin. Masters thesis. Cornell University

Van Der Schalie H (1974) Aswan Dam revisited. Environ Sci Policy Sustain Dev 16:18–20. https://doi.org/10.1080/00139157.1974.9928525

Whittington D, Waterbury J, Jeuland M (2015) The Grand Renaissance Dam and prospects for cooperation on the Eastern Nile. Water Policy 16(4):595–608

Chapter 7
Grand Ethiopian Renaissance Dam Reservoir Filling

Abstract The Grand Ethiopian Renaissance Dam (GERD) is expected to create a reservoir of 73 bcm (billion cubic meter) covering an area of 1883 km^2 stretching 246 km upstream. The reservoir will be half the size of the upstream Lake Tana. After the surprise launching of the construction of GERD, the controversy around GERD quickly evolved from its engineering design and construction aspect to the impact of filling on the Sudan and Egypt. One of the controversies with Egypt is the number of years for initial reservoir filling, as shorter filling time requires more flow reduction and higher investment return from the dam. Longer filling time requires lower flow reduction and lower investment return from the dam. In this chapter, reported and synthetically generated stream flow data were used to estimate required period of time for initial reservoir filling. Results suggest that eight to nine years is likely required to fill the reservoir with 20% of the annual flow of each year held back. With a historical average Nile annual flow of 84 bcm at Aswan, the 132 bcm volume of the High Aswan Dam took six years (1971–1976) to reach full capacity with full river flow. Historical average Nile annual flow at Aswan is 84 bcm.

Keywords Grand Ethiopian Renaissance dam · Ethiopia · Blue Nile River Reservoir filling · Dam operation · Ethiopia · Sudan · Egypt

7.1 Introduction

The Grand Ethiopian Renaissance Dam (GERD) is the first major dam on the Blue Nile (Abay) River in Ethiopia while there are several dams on the Nile in the Sudan and Egypt (Fig. 7.1). Along the Blue Nile, there are several potential hydropower sites that can generate more power with a series of dams. The GERD is located at the last possible dam site before the river crosses to the Sudan. GERD is the first major dam on the main Blue Nile River of Ethiopia.

There are over 25 dams and major water control infrastructures on the Nile River. The Sennar and Roseires dams in Sudan are on the Blue Nile. The hydrology of the system is that upstream generates flow and downstream expects to receive flow with

© Springer International Publishing AG, part of Springer Nature 2019   97
W. Abtew and S. B. Dessu, *The Grand Ethiopian Renaissance Dam on the Blue Nile*, Springer Geography, https://doi.org/10.1007/978-3-319-97094-3_7

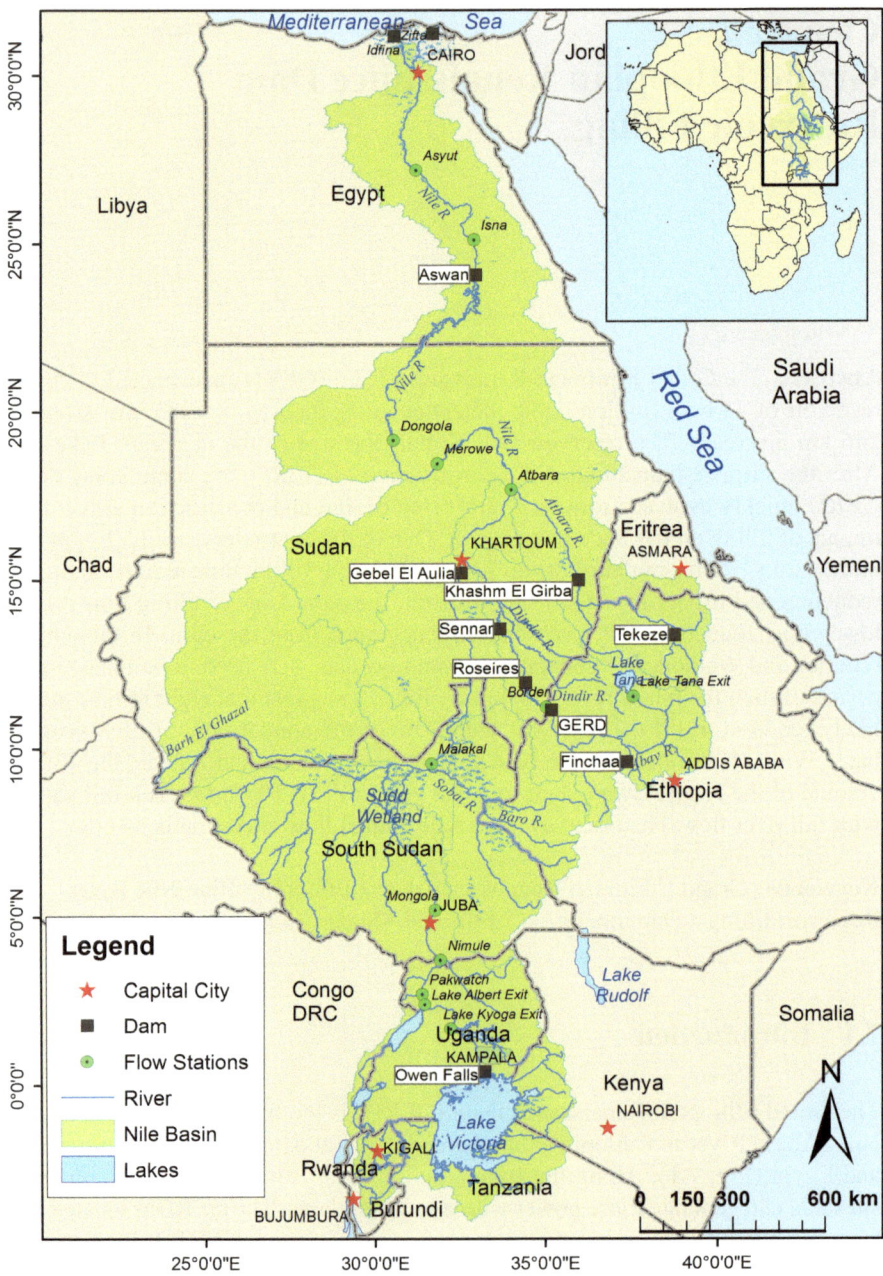

Fig. 7.1 Location of major Dams in the Nile Basin

little to none reduction. The challenge for filling and operating dams in the Blue Nile basin is downstream water demand.

The High Aswan Dam (HAD) in Egypt is the largest dam in the Nile River Basin with a reservoir capacity of 132 bcm. Upon completion in 1970, the HAD was intended for sustainable irrigation development, hydropower and navigation improvement (Abu-Zeid and El-Shibini 2010). The dam has been credited to save Egypt from extreme floods and droughts that devastated the upstream riparian nations (Abu-Zeid and El-Shibini 2010; Strzepek et al. 2008). The operation of the HAD is regulation of Nile flood water for irrigation, power generation and flood control with little obligation to downstream water demand which is environmental demand of the delta. The filling and operation was not a source of conflict because there was no upstream demand.

During the construction of the High Aswan Dam in Egypt with the support of the Soviet Union, the United States Bureau of Reclamation (USBR) studied potential hydropower dam sites on the Blue Nile in Ethiopia from 1956 to 1964. In 1964, it offered four proposed dam sites, Karadobi, Mabil, Mendia and Border (Fig. 7.2). The current Ethiopian Government selected the Boarder site and began construction of the GERD in April 2011 for the sole purpose of hydroelectric power generation. The sole purpose of the dam was announced to be hydroelectric power for consumption in Ethiopia and export to Egypt and Sudan.

The Grand Ethiopian Renaissance Dam (GERD) is expected to create a reservoir of 73 bcm (billion cubic meter) covering an area of 1883 km^2 stretching 246 km upstream. The reservoir will be half the size of the upstream Lake Tana (Fig. 7.3). Dam site selection matters in the dam filling and operation scenarios which are subject to the influence of upstream and downstream climate. The location provided the maximum potential to tap the available flow of all the Blue Nile tributaries in Ethiopia. However, the relatively lower relief topography has resulted in large surface area and combination of dams. The Blue Nile River flow has high variation with annual flow volume of 20.6 bcm (1913) to 79 bcm (1909) reported with decadal mean of 37.9 from 1978 to 1987 (Conway 2000). The GERD filling period climatic condition, dry or wet, will determine the number of years of filling and the potential cooperation or conflict both upstream and downstream of the dam.

7.2 Flow Characteristics and Flow Data

Flow characteristics and sediment load are important parameters in dam site selection and design. Blue Nile River flow seasonal distribution varies with rainfall distribution and flood travel time. Average annual flow of Blue Nile (Abay) at Ethio-Sudan boarder is approximately 50 bcm. Flow temporal variation patterns rainfall variation except river flow lags with distance from the runoff generating sub-basins. Flood flow at Ethio-Sudan border and Sennar dam downstream is from July to October while it is August to November at Aswan dam. Ethio-Sudan border and Sennar flows are from the Blue Nile while Aswan receives additional flows from the White Nile, Sobat and

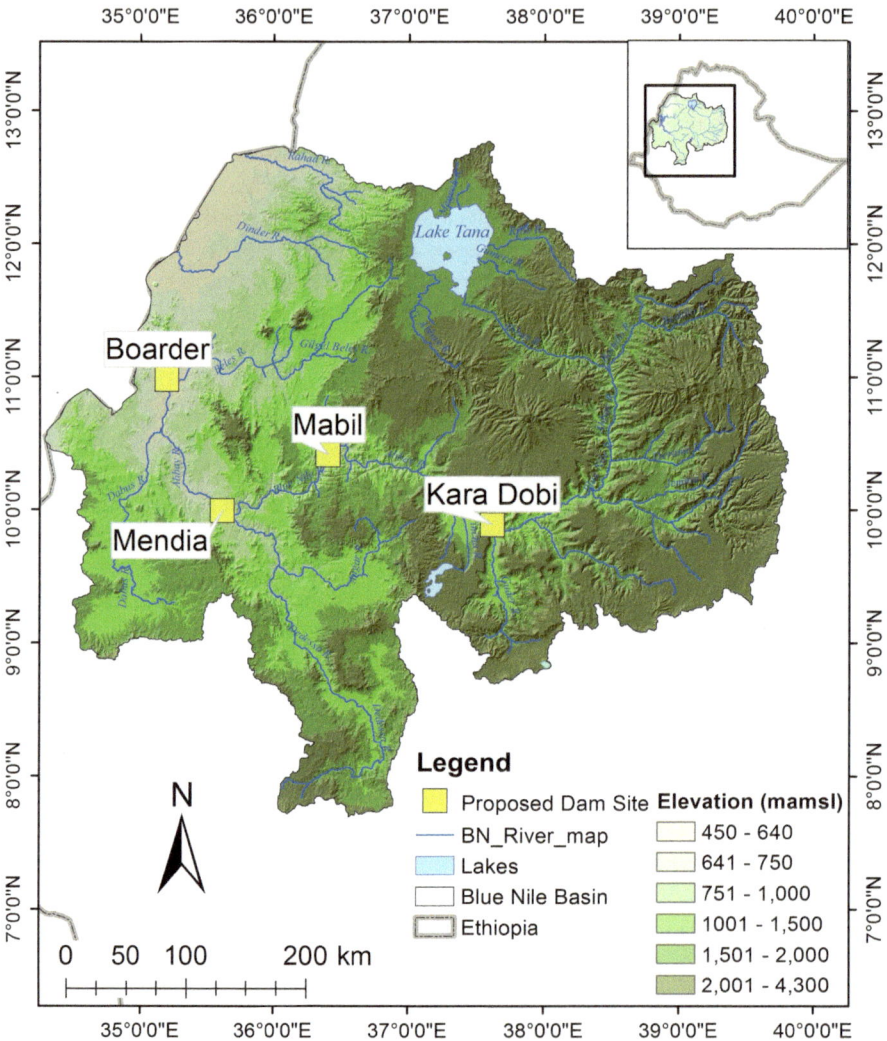

Fig. 7.2   1964 USBR proposed hydroelectric dam sites on the Blue Nile River Basin, Ethiopia

Atbara. July through October flow at the dam site is 80% of the annual flow (Abtew and Melesse 2014). Figure 7.4 depict temporal characteristics of Blue Nile river flow at border from multiple data sets. Blue Nile River design flow characteristics at the dam site are depicted in Table 7.1. Mean monthly flow at Sennar dam and border comparing multiple data sources are shown in Fig. 7.5.

Published annual flow data estimates for the Blue Nile River at the border with Sudan and at Roseires dam downstream in Sudan vary with reporting, 47.1, 48.7, 48, 41.7 bcm (Mulat et al. 2014; FAO 2018; Whittington et al. 2015; Keith et al. 2014). Another estimate of average annual flow at Roseires from graphic report (1912–1991)

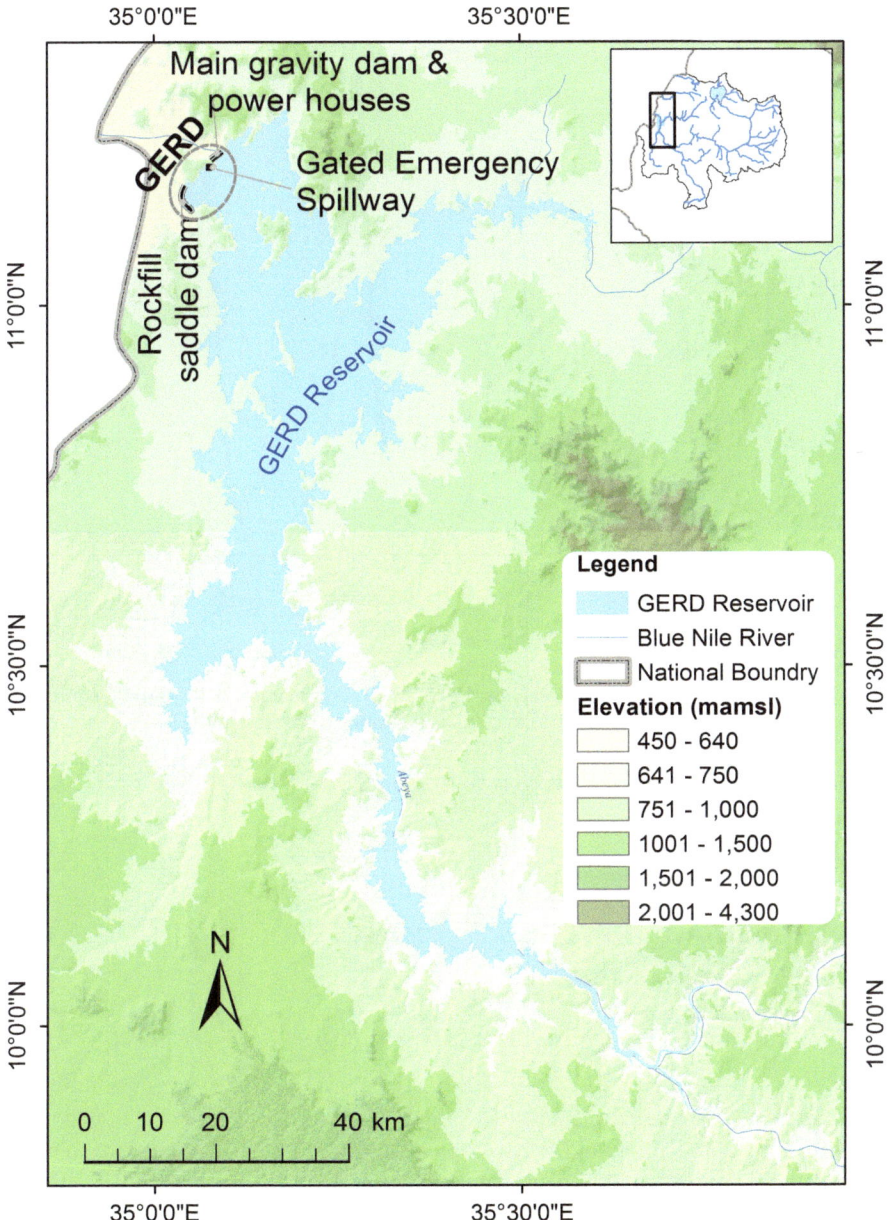

Fig. 7.3 GERD dam location and extent of reservoir at full supply level

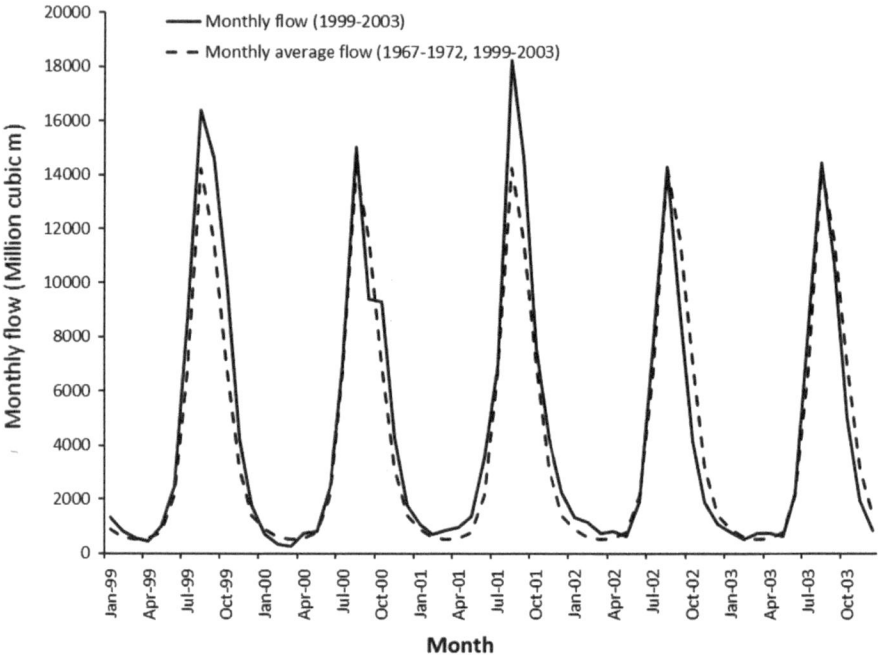

Fig. 7.4 Monthly measured flow (solid line, 1999–2003) and average monthly flow at border (dash line, 1967–1972; 1999–2003) overlaid to compare variability over the two periods

Table 7.1 Blue Nile River design flows at GERD dam site (IPoE 2013)

River characteristics	
Mean flow	$1547 \text{ m}^3 \text{ s}^{-1}$
10-yr flow	$14{,}700 \text{ m}^3 \text{ s}^{-1}$
10,000-yr flow at 641.8 m asl	$26{,}860 \text{ m}^3 \text{ s}^{-1}$
Probable Maximum Flood at 642.9 m asl	$38{,}750 \text{ m}^3 \text{ s}^{-1}$
Mean annual sediment yield	207 million m^3

by Tesemma (2009) is estimated as 50 bcm. Five years flow data at the Sudan border (GERD site) from Ethiopian Meteorological Agency (1999–2003) has mean annual flow of 53.7 bcm. The combined annual average flow estimate at GERD site from the above two sources (1967–1972; 1999–2003) is 50 bcm.

For the purpose of reservoir operation, simulation ensembles of synthetic data were generated from statistics and distribution of reported flows. Figure 7.6 depict annual flow variation extracted from various sources and median of generated synthetic annual flow data.

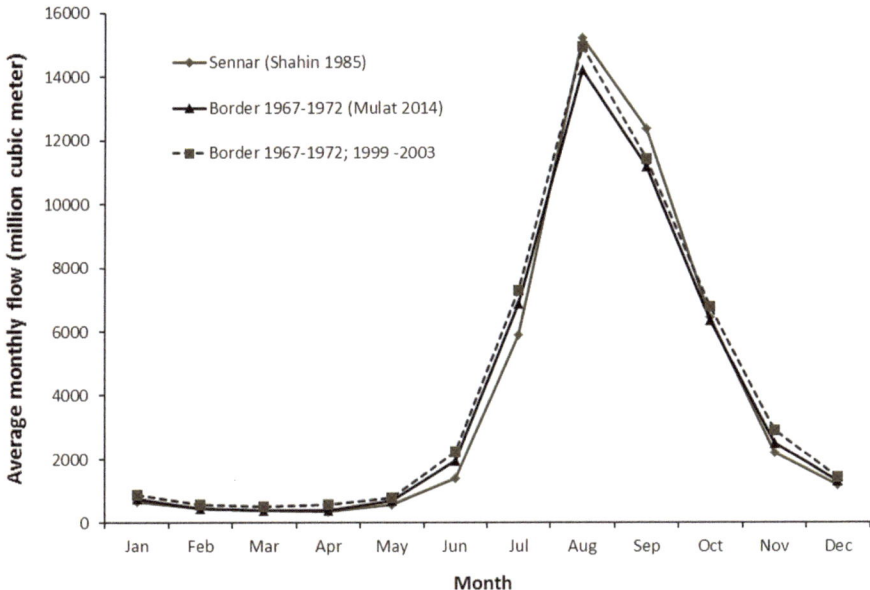

Fig. 7.5 Mean monthly flow at border and Sennar dam

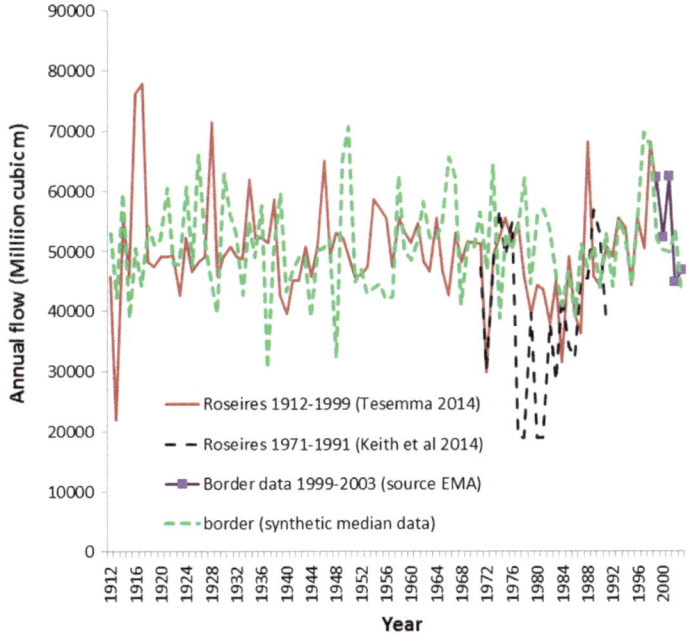

Fig. 7.6 Annual Blue Nile River flow estimates

7.3   Synthetic Flow Data Generation to Expand Data Range

Synthetic hydrometeorology data generation for hydrologic systems simulation is widely applied (Abtew et al. 1990). Multiple ensembles of synthetic flow data generated from statistical characteristics of observed or reported data can be applied to systems as reservoir operations and estimating reservoir filling period. Results with certain confidence level can be generated from multiple ensembles of simulations. Hundred by hundred (10,000 years) of synthetic flow data was generated based on the statistics of 11 years (1967–1972; 1999–2003) flow data at GERD site (border) using normal probability distribution (Eq. 7.1).

$$X = \mu + Z\sigma \tag{7.1}$$

where X is synthetic annual flow, μ (50.38 bcm) is observed expected annual flow, Z is probability corresponding to the normal distribution and σ (8.35 bcm) is standard deviation of the observed or reported annual flow. The median of synthetic annual flow data is depicted in Fig. 7.6 along with other reported annual flows. Figure 7.7 depicts exceedance probabilities of 1912–1999 border flow data derived from Tesemma (2009) and five sets of synthetic annual flows for the purpose of demonstrating the comparability of synthetic flow data with reported data. Both the reported and synthetic data were applied in GERD reservoir filling simulation with similar outputs.

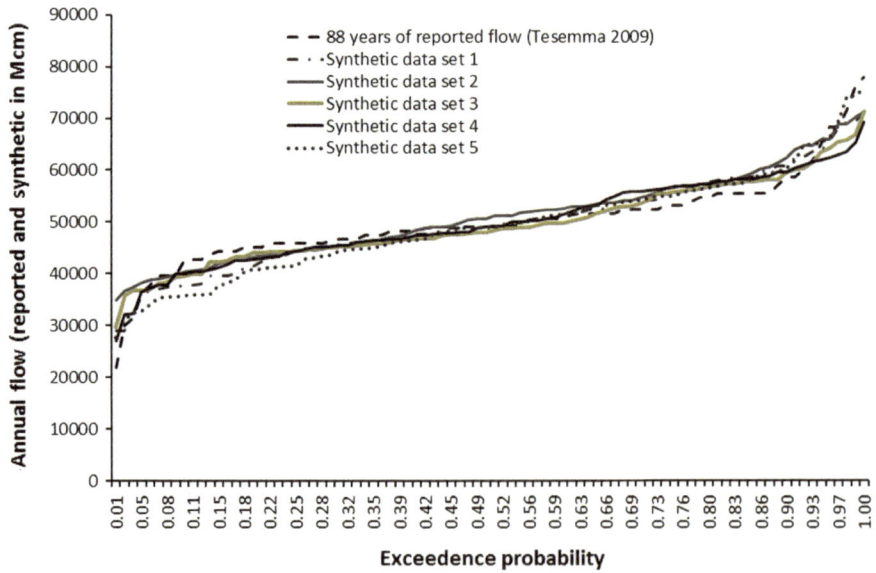

Fig. 7.7 Comparison of reported and synthetic Blue Nile annual flows at GERD site with exceedance probabilities

7.4 Evaporation and Seepage Losses

The estimated annual water loss due to evaporation and seepage from High Aswan Dam (HAD) ranges from 24 to 10 bcm per year (Abu-Zeid and El-Shibini 2010). HAD water loss due to evaporation and seepage is estimated as 14% of inflow, 12.5 bcm assuming surface area of reservoir as 5000 km^2 with water level of 170 to 175 m asl. Seepage is estimated as 5 bcm (Hussein 1981). In terms of depth of water loss over the average reservoir surface area, evaporation is 1.5 m and seepage is 1 m. This evaporation loss is far lower than what is generally accepted for the area of the dam, 2.19 m (Sadek et al. 1997), 2.7 m (Omar and El-Bakry 1981), 2.47 m (Shahin 1985) and 2.4 m (Hassen 2013). Evaporation and seepage losses are major factors in the Nile basin. Since rainfall over the HAD reservoir is not considerable, the reservoir is characterized by its high system loss.

At GERD, the difference between rainfall on the reservoir and evaporation is estimated as 0.73 m per annum from reservoir surface area at full service level (Mulat et al 2014). Potential evaporation or free water evaporation for the area is estimated as 1772 mm (Asosa) as shown in Chap. 4 of this book. Rainfall in the area comes out 1042 mm as the difference between rainfall and evaporation to remain at 0.73 m over the full service level reservoir area. Seepage losses for the GERD can be estimated as 2.1 mm per day for clay sandy or sandy clay soils (Peppersack 2015).

7.5 Sediment

Erosion of Blue Nile basin soils is self-evident from historical deposition of fertile soils of the Nile Delta in Egypt. Erosion in the Upper Blue Nile basin is estimated as 131 million t yr^{-1} in mass (Betrie et al. 2011) and 207 million m^3 in volume (IPoE 2013). Soil loss in Upper Blue Nile basin are due to deforestation for fuel and agricultural land clearing exacerbated by steep topography, poor agricultural practices, lack of resource management, and unfavourable land tenure system. Case studies of current soil loses are documented (Tammo et al. 2014; Gessesse 2014; Engdayehu et al. 2016). Population pressure and the inability to transition from biomass to other forms of energy for domestic use are aggravating the problem. Loss of storage in dams in the Sudan and the HAD in Egypt due to sedimentation has been documented in some cases raising dam heights to mitigate the problem. Due to its location, GERD will face higher sedimentation problems reducing storage and hydropower head and the useful life of the dam. Bottom outlets effectiveness and operation to flush out high-sediment load in the wet season could mitigate the sedimentation problem. Taking the routing time into account, the first two months of the rainy season will bring significant load of sediment.

7.6   GERD Filling

Literature on GERD filling has started coming out with diverse conclusions (Cascão and Nicol 2016; McCartney and Girma 2012; Strzepek et al. 2008; Keith et al. 2014; Wheeler et al. 2016; King 2013; Zhang et al. 2015; Tesemma et al. 2014). There is no publicly declared policy on filling the GERD. Based on evaporation losses, climate variability and climate change, the filling policy has downstream implications. In a study, rainfall-runoff, flow routing and hydropower simulations were applied at monthly time step for the period 2011–2060. Due to tributary conditions and climatic variability, it was reported GERD filling rate and downstream flows has non-linear relationship. The study result showed a filling policy impounding 10% of monthly flow behind GERD reduces 6% flow and a filling policy of 25% reduces 14% flow at Lake Nasser (High Aswan Dam) in Egypt for the first 5 years. The study concluded the impact on Geziera irrigation of Sudan will be more. These results are subject to climate change. A second filling policy, filling the GERD only when flow is above the historical average, will cause 7% flow reduction at Lake Nasser. The results show high variability (Zhang et al. 2015).

A study that applied the RiverWare model to simulate flow through the reaches of the Nile has reported results on hydropower and flow impact of GERD filling using multiple traces of hydrology (Tesemma et al. 2014). This study run scenarios of 5, 6, and 7 years filling period and produced impacts after 13 years with filling starting with Aswan High Dam initial stage of 173 m asl (Table 7.2).

A study from US military Academy at West Point applied 33 GCM models to foresee precipitation and climate change in the Nile basin (Keith et al. 2014). The study also applied system dynamics modelling to predict political, economic and social impacts of predicted Nile stream flows. The climate change modelling predicted both precipitation and temperature increasing with precipitation increase in flow source areas and temperature increasing more in the Sudan and Egypt. With this background, the impact of GERD filling was evaluated. Historical flow data from 1994 to 2012 coupled with theoretical distributions and climate change, simulated flows were generated for the period 2014–2100 in 30-year intervals. The estimated conclusions from the study on GERD filling are shown in Table 7.3. Generally short filling period is expected to create system shock and long filling period could create long-term stress.

Table 7.2 Average impact of 5, 6, 7 year filling period for GERD (extracted from Tesemma et al. 2014)

Country	Hydropower (%)	Flow (%)	Location
Ethiopia	400		
Sudan	14.47	−3.30	El-diem (Ethio border)
Egypt		−8.02	Aswan dam inflow
Egypt	−8.34	−3.01	Aswan dam outflow

Table 7.3 Estimated impacts of GERD filling years on Blue Nile flow (extracted from Keith et al. 2014)

GERD fill rate (% of full supply dam volume)	Years to fill	Estimated percent reduction in Blue Nile flow
5	20	−4.22
10	10	−8.77
15	7	−13.15
20	5	−16.18
25	4	−20.72

A study was conducted assessing GERD reservoir filling policies and climate change on the number of years required to fill the dam to FSL (Full Supply Level), (King 2013). Hydropower and reservoir model was applied for the period 2011 to 2060 with the following assumptions. Initial construction of GERD assumed to be 2011 to 2014. Final construction to be from 2014 to 2017 where, during this period, 15% of FSL is filled and partial power generation is assumed. Results for climate changes of −20, −10, +10, +20% rainfall changes and no change were applied to different dam filling policies. These policies were, holding back, 5, 10, and 25% of any monthly flow; and holding back in excess of historical average stream flow (HASF) for the month or holding back monthly flows in excess of 90% of HASF. Holding back 5% of monthly flow with no change in climate did not fill the reservoir by 2060. Simulation results for the different filling policies and climate change scenarios show the number of years since 2011 required to fill reservoir to FSL (Table 7.4). The study assumes partial filling between 2014 and 2017. Since the completion of the reservoir is delayed, the reference years may not matter.

Table 7.4 GERD filling years for various climate change scenarios and filling policies (extracted from King 2013)

Precipitation trend	Filling years for four fractional monthly flow impoundment			
	10% of Flow	25% of Flow	Excess of HASF	Excess of 90% HASF
−20	32		>50	21
−10	26		26	16
0	24	12	19	13
10	22	12	16	13
20	21	11	13	12

7.7 Simulation of GERD Initial Filling

Factors affecting the rate of initial filling of the GERD are Blue Nile river flow rates during the period of filling where droughts will elongate the filling period and wet conditions shorten the period. The impact of drought on stream flow is two-fold: reduced rainfall input and increased evaporation losses. The next major factor in determining the filling period is the rate at which the dam is filled as percentage or fraction of the incoming steam flow. Holding back 5% of the flow didn't fill the reservoir by 2060 (King 2013). Holding back higher percentage of the flow is required to shorten filling time and make the project economically viable but downstream impact is expected. Another factor in determining the filling period is the rate of seepage loss from the new reservoir where high seepage loss will extend the filling time and vice versa.

In this chapter, reservoir operation simulation is applied to estimate filling period using annual stream flow data with a fraction of the flow retained in the reservoir. Water surface area changes with time during the filling period. Figure 7.8 depicts the relationship of storage volume and surface area as expressed by Eq. 7.2. Figure 7.9 illustrates the change in water surface area during filling.

$$A_t = -0.137 * V_t^2 + 35.904V_t + 52.5563 \tag{7.2}$$

where A_t is reservoir water surface area (km^2) and V_t is reservoir storage volume (bcm) for year t.

In the absence of monthly stream flow data, annual flow data is used for dam filling simulation with the understanding that percentage withholding is applied uniformly throughout the year reflecting monthly and seasonal flow variations.

The assumption is that the reservoir operation is capable of releasing each years' flow retaining back fraction of flow determined by the filling policy without hydraulic limitations on annual basis. Day to day or month to month flows and operations may

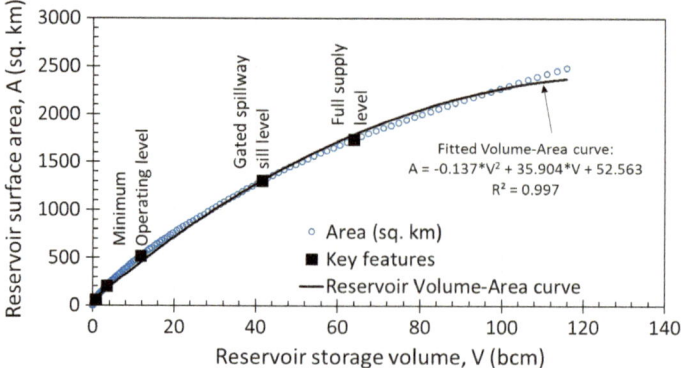

Fig. 7.8 Reservoir volume and surface area relationship

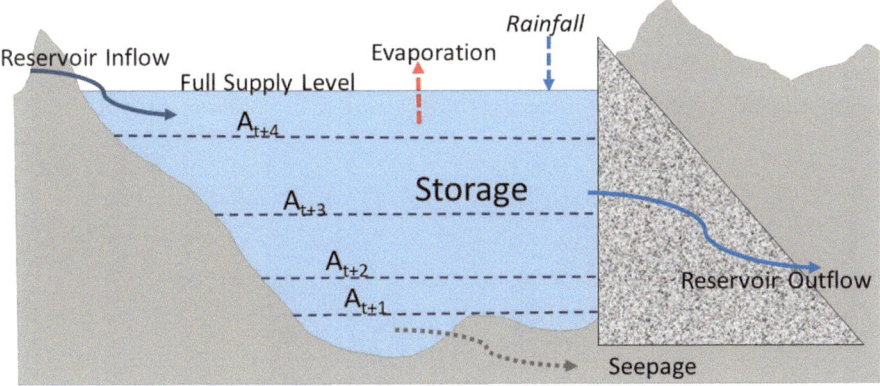

Fig. 7.9 Schematic model of reservoir surface area changes with inflows and outflows during filling

vary but result should be close to the filling policy by the end of the year. The filling policy in this study is a predetermined fraction of flow, F, retained in the reservoir regardless of flow rate.

$$V_t = V_{t-1} + Q_t * F - SP_t + (R - E)_t \text{ for } V_{t-1} \geq 0 \qquad (7.3)$$

where Q_t is the annual stream flow, F is fraction of flow retained (filling policy) SP_t is reservoir seepage loss that varies with water surface area, E is evaporation and R is rainfall for the year where the difference varies with change in water surface area until reservoir is filled. Although seepage varies with head or water level, a single value is used from seepage standards for ponds and reservoirs referred earlier.

To overcome limitation in flow data, ensembles of 100-year sequence synthetic flow data was generated using statistics from various data sources. Normal distribution was used for generating synthetic annual flow data at the GERD site. Synthetic data provide multiple ensembles of flow data to represent several possible sequences of annual flows. Simulation of reservoir filling with these large set of data provides the opportunity to estimate filling years with statistical probabilities. Synthetic data was generated based on statistics of different data sources and data periods to evaluate data source impact on reservoir filling simulation. Table 7.5a and b depict statistics of reported and synthetic flow data that were used in reservoir filling simulation to estimate number of years for initial reservoir filling. An annual average flow of the Blue Nile (Abay) has been reported as 54.4 bcm by Berhanu et al. (2014).

In this analysis, filling policy of less than 20% flow retention took long period that may affect the structural integrity and the economic feasibility of the dam. Therefore, 20% was used as the minimum percentage of flow retention to estimate number of years required for filling without making the dam an economic loss with extended filling period. The simulation assumed dam operation can retain 20% of the annual incoming flow. The retention can be uniform or seasonally varied as long as the annual retained volume approaches 20%. Simulation of reservoir filling was

Table 7.5 Statistics of reported and synthetic annual flow data and parameters used in reservoir filling simulation

a

Flow statistics	Tesemma (2009)	Synthetic (mean of 1000 years)
	bcm	bcm
Q_{avg}	50.01	50.02
Q_{std}	8.57	8.53
Q_{min}	21.89	29.02
Q_{max}	77.84	70.72
E–R (average)	0.73 m (Mulat et al. 2014)	
Seepage	0.779 m (sandy clay or clay sandy soil 2.134 mm/d)	

b

Flow statistics	1967–1972; 1999–2003	Synthetic (mean of 10,000 years)
	mcm	mcm
Q_{avg}	50.38	50.36
Q_{std}	8.35	8.30
Q_{min}		29.61
Q_{max}		71.14
E-R (average)	0.73 m (Mulat et al. 2014)	
Seepage	0.779 m (sandy clay or clay sandy soil 2.134 mm/d)	

performed with sets of 100-year synthetic flow data generated from two reported flow sources. Probabilities and cumulative probability of estimated reservoir filling years for filling policy of retaining 20% of the annual flows is depicted by Fig. 7.10. The conclusion from this analysis is that it is likely to take eight to nine years to fill the reservoir with 20% flow withholding.

The two synthetic flow scenarios (Fig. 7.10) suggested that a 20% annual retention will likely fill the reservoir in 8–9 years (60% probability) with smaller probability of 7 or 11 years. Wheeler et al. conducted filling simulation and reported a 7 years filling period with 40 bcm of annual flow released, referred to "agreed release" in the study, with first year initial filling to 560 m asl (Wheeler et al. 2016). This is equivalent to 20% of average flow retained. Our result (Fig. 7.11) is close to this study which applied detailed modelling flow along the Nile through several dams. These studies can be used as reference in developing filling policy. Table 7.6 depicts extracted results of GERD filling study for various amounts of annual releases (Wheeler et al. 2016) and percent retained and released added. According to a detailed study of GERD impact on the Upper Egypt irrigation and other water needs, the reported maximum allowable reduction of flow for GERD impoundment is 5–15% (Abdehaleem and Hatat 2015).

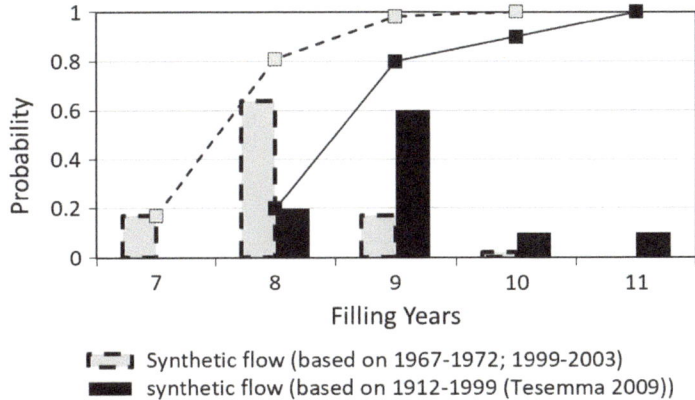

Fig. 7.10 Number of reservoir filling years and probabilities

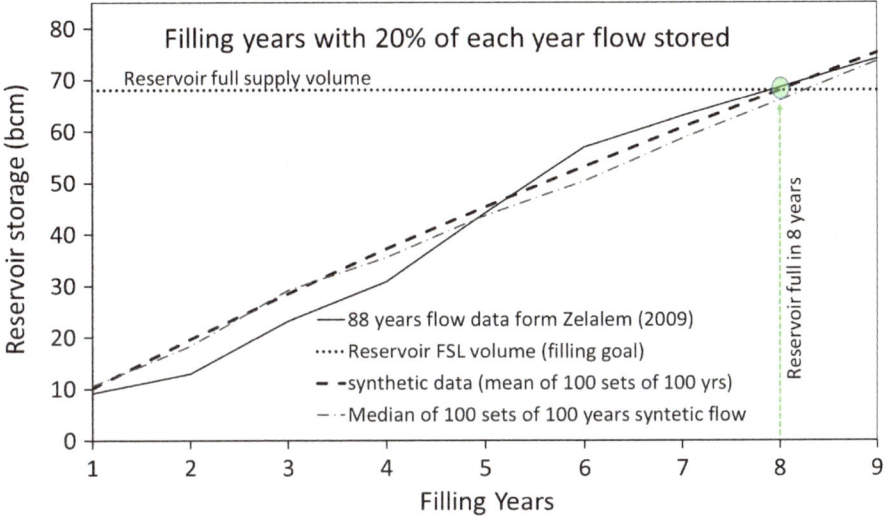

Fig. 7.11 GERD filling simulation with various estimates of annual Blue Nile River flow

7.8 Summary

The economic value of hydropower and the downstream impact of GERD will not be any less than the HAD. Besides the geo-political implications, GERD is vital to the energy starved economy of Ethiopia. The GERD will continue to be controversial with downstream countries and internally within Ethiopia with respect to water rights. The initial filling of GERD will be contentious due to the loss of economic return associated with extended filling period. Long filling period adds to the question of the economic viability and structural integrity of the dam but lowers downstream

Table 7.6 Years to fill the GERD with various annual releases ([a]extracted from Wheeler et al. 2016)

Annual release (bcm)[a]	Mean filling years[a]	Percent of 50 bcm released (retained)
0	3	0 (100)
25	4.5	50 (50)
30	5	60 (40)
35	5.5	70 (30)
40	**7**	**80 (20)**
45	14	90 (10)

flow reduction. The second point of contention is dam operation for conflicting objectives of not impacting downstream and optimal power generation. We presented the possible flow reduction due to filling and the implication on downstream storages. We also cited multiple scenarios of initial reservoir filling and compared with the corresponding filling period of the HAD. Despite the course of filling schedule, the riparian countries can benefit from negotiation on the filling of GERD to reduce unintended consequences.

References

Abdehaleem FS, Hatat EY (2015) Impacts of Grand Ethiopian Renaissance Dam on different water usage in Upper Egypt. Br J Appl Sci Technol 8:461–483
Abtew W, Melesse AM (2014) Chapter 2: The Nile River Basin. In: Melesse AM, Abtew W, Setegn S (eds) Nile River Basin: ecohydrological challenges, climate change and hydropolitics. Springer, New York
Abtew W, Moras RG, Campbell KL (1990) Synthetic precipitation data generation. In: Proceedings of the 12th annual conference on computers and industrial engineering. Computers and industrial engineering 19:582–586
Abu-Zeid MA, El-Shibini FZ (2010) Egypt's high Aswan Dam. Int J Water Resour Dev 13:209–218. https://doi.org/10.1080/07900629749836
Berhanu B, Seleshi Y, Melesse AMR (2014) Surface water and groundwater resources of Ethiopia: potentials and challenges of water resources development. In: Melesse AM, Abtew W, Setegn S (eds) Nile River Basin ecohydrological challenges, climate change and hydropolitics. Springer, New York
Betrie GD, Mohamed YA, van Griensven A, Srinivasan R (2011) Sediment management modelling in the Blue Nile basin using SWAT model. Hydrol Earth Syst Sci 15(3):807–818
Cascão AE, Nicol A (2016) GERD: new norms of cooperation in the Nile Basin? Water Int 41:550–573. https://doi.org/10.1080/02508060.2016.1180763
Conway D (2000) The climate and hydrology of the Upper Blue Nile River Geogr J 49–62
Engdayehu G, Fisseha G, Mekonnen M, Melesse AM (2016) Evaluation of technical standards of physical soil and water conservation practices and their role in soil loss reduction: the case of Debre Mewi watershed, north-west Ethiopia. In: Melesse AM, Abtew W (eds) Landscape dynamics, soils and hydrological. Springer, New York processes in varied climates. Springer, New York

FAO (Accessed 24 March 2018) Hydrologic regime in the Nile Basin. http://www.fao.org/docrep/015/an530e/an530e.pdf

Gessesse GD (2014) Assessment of soil erosion in the Blue Nile basin. In: Melesse AM, Abtew W, Setegn S (eds) Nile River Basin ecohydrological challenges, climate change and hydropolitics. Springer, New York

Hassen M (2013) Evaporation estimation for Lake Nasser based on remote sensing technology. Ain Shams Eng J 4(4):593–304

Hussein MF (1981) Dams, people and development: the high Aswan Dam case. Pergamon Press Inc., New York

IPoE (2013) International panel of experts on grand Ethiopian Renaissance Dam Project. Final Report, Addis Ababa, Ethiopia

Keith B, Epp K, Houghton M, Lee J, Mayville R (2014) Water as a conflict driver: estimating the effects of climate change and hydroelectric dam diversion on the Nile River stream flow during the 21st century. Center for Nation Reconstruction and Capacity Development, United States Military Academy, West Point, New York

King AM (2013) An assessment of reservoir filling policies under changing climate for Ethiopia's Grand Renaissance Dam. Masters Thesis, Drexel University

McCartney MP, Girma MM (2012) Evaluating the downstream implications of planned water resource development in the Ethiopian portion of the Blue Nile River. Water Int 37:362–379

Mulat AG, Moges SA, Ibrahim Y (2014) Chapter 27 Impact and benefit study of Grand Ethiopian Renaissance Dam (GERD) during impounding and operation phases on downstream structures in the Eastern Nile. In: Melesse AM, Abtew W, Setegn S (eds) Nile River Basin ecohydrological challenges, climate change and hydropolitics. Springer, New York

Omar MH, El-Bakry MM (1981) Estimation of evaporation from the lake of the High Aswan Dam (Lake Nasser) based on measurements over the lake. Agric Meteorol 23:293–308

Peppersack J (2015) Seepage loss standards for ponds and reservoirs. Application Processing Memorandum No. 76. Idaho department of Water Resources. Idaho

Sadek MF, Shahin MM, Stigter CJ (1997) Evaporation from the reservoir of the High Aswan Dam, Egypt: a new comparison of relevant methods with limited data. Theoret Appl Climatol 56(1):57–66

Shahin M (1985) Hydrology of the Nile basin. Elsevier, New York

Strzepek KM, Yohe GW, Tol RSJ, Rosegrant MW (2008) The value of the high Aswan Dam to the Egyptian economy. Ecol Econ 66:117–126. https://doi.org/10.1016/j.ecolecon.2007.08.019

Tammo SS, Tilahun SA, Tesemma ZK, Tebebu TY, Moges M, Zimale FA, Worqlul AW, Alemu ML, Ayana EK, Mohamed YA (2014) Soil erosion in the Blue Nile basin: trends and challenges. In: Melesse AM, Abtew W, Setegn S (eds) Nile River Basin ecohydrological challenges, climate change and hydropolitics. Springer, New York

Tesemma ZK (2009) Long term hydrologic trends in the Nile basin. Masters thesis, Cornell University

Tesemma ZK, Mersha A, Wheeler K (2014) Reservoir filling options assessment for the Grand Ethiopian Renaissance Dam using a probabilistic approach. Grand Ethiopian Renaissance Dam Workshop. Abdul Latif Jamal World Water and Food Security Lab, MIT, MA, 13–14 Nov 2014

Wheeler KG, Basher M, Mekonnen ZT, Eltoum SO, Mersha A, Abdo GM (2016) Cooperative filling approaches for the Grand Ethiopian Renaissance Dam. Water Int 41:611–634. https://doi.org/10.1080/02508060.2016.1177698

Whittington D, Waterbury J, Jeuland M (2015) The Grand Renaissance Dam and prospects for cooperation on the Eastern Nile. Water Policy 16(4):595–608

Zhang Y, Block P, King A (2015) Ethiopia's Grand Renaissance Dam: implications for downstream riparian countries. J Water Resour Plann Manage 141(9):1–18

Chapter 8
Grand Ethiopian Renaissance Dam Operation and Upstream-Downstream Water Rights

Abstract Hydropower dam operation and economic return is subject to climatic variation and downstream and upstream water demand related constraints. The Grand Ethiopian Renaissance Dam (GERD) design installed power generation capacity is 6000 MW. The average flow rate of the Blue Nile will generate a third of the installed power with average annual energy production of 15,700 GWH. Optimal operation of the GERD will be critical to market sustainable power output as a result of immediate downstream water demand, long-term upstream water demand and climatic variation. Long construction period, long initial reservoir filling period, and refilling period after drought, puts to test both the reliability of power generation, and cooperation to synchronize operation of GERD and downstream dams. Regulation schedule of the dam is dependent on sets of priorities. Simulation of reservoir operation with two minimum flows, 1500 and 1600 $m^3\ s^{-1}$ was performed and results presented. Operation is sensitive to change in minimum flow and annual flow variation.

Keywords Dam operation · Nile River Basin · Grand Ethiopian Renaissance dam · Ethiopia · Egypt · Sudan

8.1 Introduction

There are two kinds of dams for hydropower generation. These are run-of-the-river hydropower generation and hydropower generation with dam and storage reservoir. Run-of-the-river requires none too little storage and operates without affecting downstream flow. Power generation rate fluctuates with stream flow. An example of this is the old Tis Abay I hydropower station (1964) on the Blue Nile, downstream of Lake Tana in Ethiopia. A run-of-the-river hydropower generation with some form of storage is Tisa Abay II hydropower station (2001) which benefits from some storage management of Lake Tana, 35 km upstream. The building of the Chara Chara weir on Lake Tana (1996), at the outflow to the Blue Nile River, provides flow rate regulation when the lake level is below 1788 m asl, the elevation of the spillway.

Dam and reservoir hydropower requires storage of water and optimal regulation schedule and operation for maximum and dependable power production. Hydropower generation is dependent on water level in the reservoir and discharge rate. As a result, timing and magnitude of downstream flow is changed from pre-dam periods. Additional losses associated with reservoirs are evaporation and seepage. Evaporation is more important if the site is in a dry region such as the High Aswan Dam (HAD) (AbAbu-Zeid and El-Shibini 2010) where difference between evaporation and rainfall is high. Dam site selection is important for such dams as site topography determines area of reservoir, volume, cost of construction, risk, safety, security, evaporation and seepage losses, and potential power generation. Figure 8.1 depicts location of the GERD and other major dams on the Nile River. GERD is the first major dam in the main Blue Nile (Abay) river in Ethiopia.

Hydropower systems are susceptible to droughts where downstream demand can compound challenges. For example, Venezuela faced a series power shortage at its major reservoir at Guri dam which receded to 5 m off dead pool due to the 2015/2016 El Niño, risking air entrapment in the turbines (The New Yorker, 17 May 2016). Likewise, Brazil faced drought from 2012 to 2015 impacting storage level in its reservoirs. Decline in storage resulted in hydropower reduction of 20% initiating increase in thermal power generation to combat the power deficit (Zambon et al. 2016). After every drought and storage loss, refilling of reservoir is necessary which becomes challenging with downstream and upstream water demand constraint.

This chapter is not intended to discuss who has water rights of the Nile waters, rather, discuss the impact of upstream and downstream water demand and use on hydropower reservoir operations. Since GERD is located at the downstream end of the Blue Nile River in Ethiopia close to the Ethio-Sudan Boarder, the term upstream in this chapter covers most of the area of the basin in Ethiopia. The downstream water right discussed in this chapter covers the extent from GERD to HAD due to the presence of the HAD with a significant storage and operational flexibility.

8.2 Upstream Water Rights and Water Demand

One of the factors that influence meeting downstream expectations at the same time generating power cost effectively are upstream water rights and water use. As the Blue Nile basin population has grown and drought and food shortage becoming frequent, demand for water has been increasing. Upstream water demand can be compounded by unfavourable climatic fluctuation in magnitude and timing of rainfall. When wide scale shift is made from rain-fed agriculture to partial irrigation or full irrigation, water use will exponentially increase in the Blue Nile basin reducing flows to GERD.

There is no explicit national water right law in Ethiopia that the authors have identified. Article 89 (5) of the Ethiopian constitution broadly declares the state holds the right for land and other natural resources that puts water resources under state control. Ad hoc local water committees and consensus based traditional framework has been handling small-scale water distribution and resolving conflicts in some

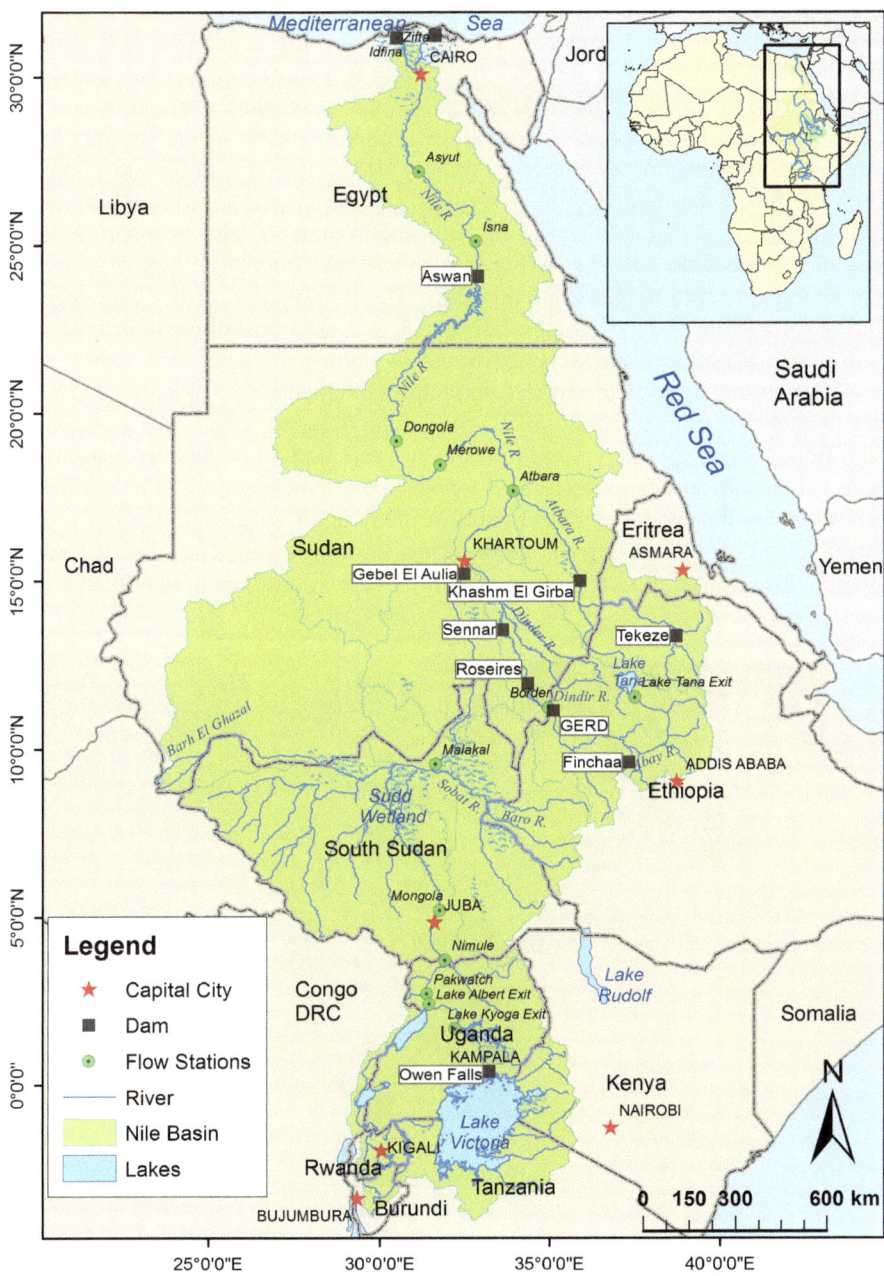

Fig. 8.1 Major dams and GERD in the Nile basin

areas. Detailed case study of Fogera and Guba Lafto District (Woreda) farmers' water conflict on small streams and efforts to develop water use guidelines and the roles of different social players, government structures, and technology (pumps) was reported by Deneke (2014). Although on small scale, this case study shows local water conflict between those who can afford pumps are diverting and drying up small streams disadvantaging downstream gravity users. The economic and population pressure forces farmers to shift to cash crops and intensive farming that usually require irrigation. Rainfall harvesting with small ponds has also become a common practice in the Blue Nile Basin (Fig. 8.2). The Lake Tana basin and other parts of the Blue Nile basin demonstrate this shift. The introduction of rice and vegetable farming are some of the examples. Large scale irrigated agriculture in the Nile basin will grow with increase in consumptive use of water.

Commercial large-scale irrigated agriculture run by the state or national and inter-national investors upstream of GERD need to be factored in reservoir management. The Belles Sugar Development Project is an example with 75,000 ha planting area and 3 sugar processing factories (Ethiopian Sugar Corporation 2015). It is located 576 km from the capital and lies in both Amhara and Benishangul Gumuz regional states. Water is diverted from the Belles River through a weir. The source of water is mainly diversion from Lake Tana which otherwise would flow to the GERD. Given

Fig. 8.2 Water harvesting pool with plastic liner in the Nile basin

the growing need for food and energy production, potential future small dams and weirs in the Blue Nile basin may include irrigation schemes that will affect the inflow to the GERD reservoir. The changing landscape of upstream consumptive water demand, water use, and other losses will need to be factored into the operation of all dams along the Nile River including GERD.

8.3 Downstream Water Rights and Water Demand

Water right claims has been part and parcel of the history of the Nile and its people dependence on the river (Degefu 2003). GERD reservoir operation determines the amount of energy production and the rate of investment returns. When water rights are assumed by both downstream and upstream entities, the operation of dams is regulated with constraint water demand from both sides of the dam. Deviation from the natural downstream flow regime corresponds to higher return from hydropower indicating the importance of full control of dam operation (Kem 2013). Jager and Smith (2008) studied dam operations with downstream environmental demands and concluded regulations bounded only by minimum flows are best for power generation than flow ramping requirements which result in limitation in meeting peak power demand. They also suggested development of regulations tailored to address unique attributes of a reservoir for optimal operation to address water balance, downstream ecological demand and optimal power generation. Figure 8.3 depicts downstream impact on reservoir operation showing relations between dam operations for optimal power generation, downstream environmental and water right demands. Operation of GERD may require synchronized operation with downstream dams and water control structures based on management agreements.

Block et al. (2007) studied the benefit-cost of construction and operation of four potential dams on the Blue Nile (Karadobi, Mabil, Mendia and Border) for hydropower and irrigation (Fig. 8.4). These sites were proposed in the Blue Nile water resource feasibility study by United States Bureau of Reclamation (USBR) in 1964. They applied hydrologic model with dynamic climate capabilities to asses benefit cost of dams on the Blue Nile for power generation and irrigation. They run simulation with the historical data and data with various frequency of El Niño and La Niña events. They reported that water policy with respect to percent of water retained is a factor in determining benefit-cost ratio. The two water policies simulated were a given percent of flow retained subject to ENSO (El Niño Southern Oscillation) frequency of occurrences. Benefit-cost ratio was reported for range of percent flow retained during hydropower dam operation with ENSO occurrences.

The impact of downstream influence on water release from GERD on reservoir water level or storage and power generation is reported by Liersch et al. (2017). This study applied global climate models and reservoir operation modelling to evaluate GERD filling time, reduction in downstream flow, and power generation, under various operation scenarios. Results of simulation of the operations shows that on the average GERD will be half full and will generate 1500 MW of power.

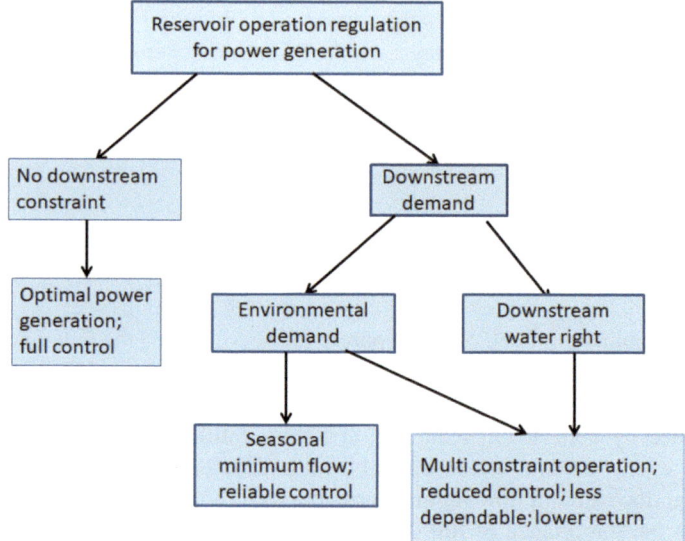

Fig. 8.3 Reservoir operation with downstream demand and power generation

8.4   Climatic Fluctuations

Climate fluctuations are major factor in water rights and water conflicts. Once a basin is in water stress, climatic fluctuations that alter the magnitude and timing of rainfall will amplify water conflicts. Too much can cause downstream flooding and deficit can create both upstream and downstream water shortage. The relationship between ENSO and regional hydrology has been determined for various regions of the world. The stronger the ENSO event the correlation is stronger resulting in excess or deficit of rainfall and runoff. Application of dynamic models with climate prediction can be applied to make flood release decisions (Li et al. 2010).

The correlation of July, August, September and October flows of the Blue Nile to Pacific sea surface temperature (SST) and the ability to forecast flow with a lead-time was reported (Eldaw et al. 2003; Abtew et al. 2009; Abtew and Melesse 2014). An analysis of relationship of Nile flow at Aswan (1872–1972) to Pacific sea surface temperature (SST), showed that 25% of the variability in flow was associated with ENSO (Eltahir 1996). It has been reported that dry years in the Blue Nile basin show a degree of association with low values of southern oscillation index (SOI), El Niño events, Conway (2000). Gissila et al. (2004) reported studies that relate El Niño events to Ethiopian droughts. ENSO and solar periods were factors both related to Nile flows in a study of decadal periodicities of the Nile River historical discharge, A.D. 622–1470, (Putter et al. 1998). Abtew et al. (2009) showed that the Upper Blue Nile basin rainfall and flows are positively related to La Niña and negatively related to El Niño. El Niño is associated with Pacific SST positive anomaly

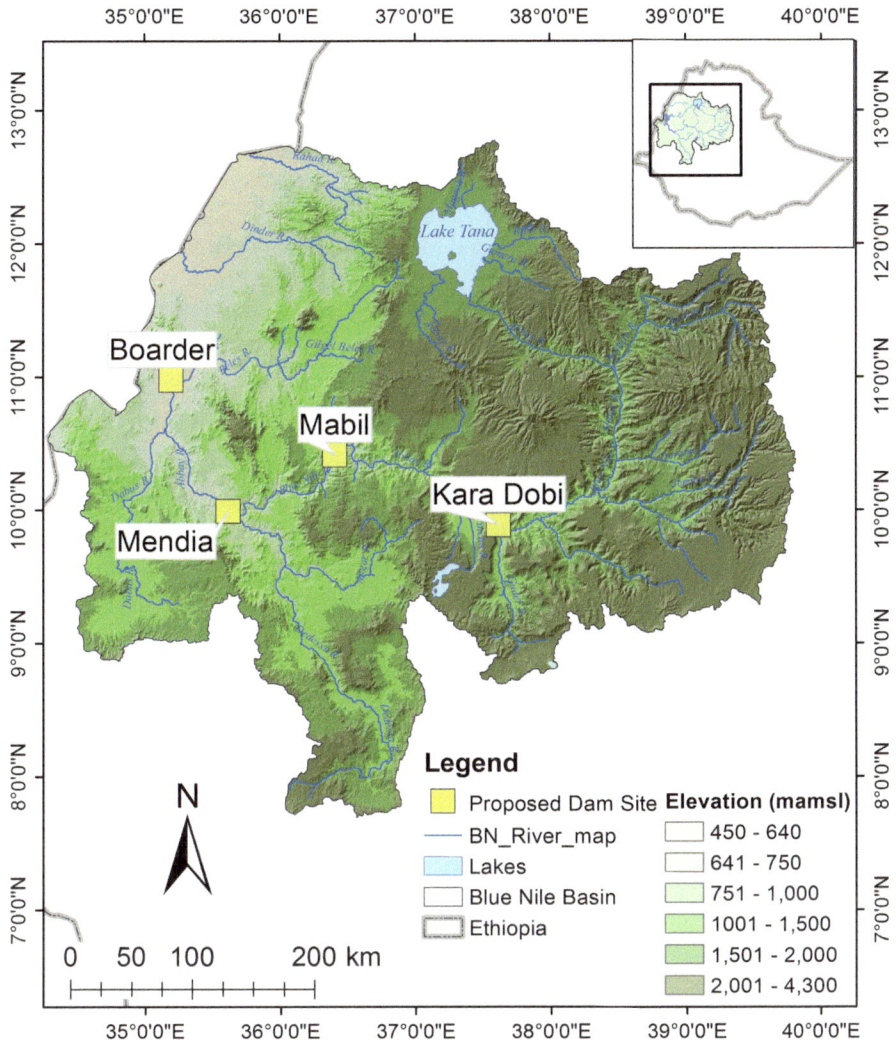

Fig. 8.4 Proposed dam site and the GERD (Border) on the Blue Nile River

and negative atmospheric pressure anomaly. La Niña is associated with Pacific SST negative anomaly and positive atmospheric pressure anomaly.

The great Ethiopian famine of 1988–1989 corresponds to the 1988 strong El Niño. The 2015–2016 strong El Niño also corresponds to extensive drought in Ethiopia. From 1960 to 2003, seven of the nine highest annual Blue Nile flows correspond to La Niña years (**1964**, **1999**, **1988**, **2000**, 2001, **1975** and 1962) with strong La Niña years shown in bold. Seven of the driest nine years occurred during El Niño years (**1994**, 1983, **1972**, **1982**, **1987**, 1990 and 2003) with strong El Niño years shown in bold (Abtew et al. 2009); Table 8.1 shows Upper Blue Nile monthly rainfall, monthly SST and cumulative SST anomalies for 1971 (wet year) and 1982 (dry year).

Cause of downstream water shortage as a result of climatic pattern induced factors can easily be mistaken with upstream water use and control. Water sharing and any water related agreement has to incorporate climate induced flow variation. The 1922 Colorado River water sharing agreement of the seven states of the United States did not include impacts of droughts. But, the 1944 Colorado River water sharing agreement between the USA and Mexico included drought impacts in water allocations (Deluca 2010). Accordingly, the operation of GERD under such extreme conditions needs to be addressed to avoid uninformed water conflict.

8.5 Reservoir Water Level and Power Generation

Hydropower generation is dependent on several factors. Water level elevation difference between the reservoir and the turbine downstream is a critical parameter in power generation referred to as head difference. The rate of water flow from the reservoir through the penstocks to the turbines is a second critical parameter for power generation. Equation 8.1 is used to calculate available power from a hydropower plant.

Table 8.1 Upper Blue Nile monthly rainfall, SST and cumulative SST anomaly for 1971 (wet year) and 1982 (dry year) (Abtew and Melesse 2014)

1971 (La Niña)				1982 (El Niño)		
Month	SST anomaly (°C)	Cumulative SST anomaly (°C)	Rainfall (mm)	SST anomaly (°C)	Cumulative SST anomaly (°C)	Rainfall (mm)
Jan	−1.42	−1.42	10.99	0.0	0.0	24.35
Feb	−1.24	−2.66	0.80	−0.1	−0.1	12.93
Mar	−1.17	−3.83	36.19	0.1	0.0	60.62
Apr	−0.84	−4.67	20.36	0.4	0.4	43.29
May	−0.65	−5.32	151.88	0.8	1.2	85.75
Jun	−0.65	−5.97	246.02	1.0	2.3	160.54
Jul	−0.48	−6.45	388.73	0.8	3.1	258.03
Aug	−0.54	−6.99	375.58	1.0	4.1	300.86
Sep	−0.62	−7.61	231.63	1.5	5.6	162.83
Oct	−0.68	−8.29	104.02	2.1	7.7	124.71
Nov	−0.73	−9.02	45.01	2.2	9.9	27.17
Dec	−0.85	−9.87	8.65	2.4	12.3	1.42
Annual			1620			1263

$$P = \frac{\eta \rho Q g h}{1,000,000} \tag{8.1}$$

where P is power in MW, η is turbine efficiency (0.85), ρ is density of water (1000 kg m^{-3}), Q flow rate (m^3 s^{-1}), g is gravitational acceleration (9.81 m s^{-1}) and h is head in m. Figure 8.5 depicts available power from GERD operation for average flow rate of 1547 m^3 s^{-1} (IPoE 2013) and design flow of 4305 m^3 s^{-1} from Salini-Impregilo (Accessed 15 October 2016), the company in charge of design and construction of GERD, and for varying head or water level. In between the two discharge rates, 2000 and 3000 m^3 s^{-1} discharge rates were also applied.

The installed power generation capacity is 6000 MW and feasible power generation of 2000 MW with average river flow rate of 1456 m^3 s^{-1} and head of 145 m was reported (International Rivers 2013). This estimate is close to what is shown in Fig. 8.5. Mulat and Moges (2014), after simulation study, reported that GERD installed power is 6000 MW but will have 33% (2000 MW) firm power generation, at 500 m asl tail water and 95% turbine efficiency. It is further stated that the peak discharge was used for designing the installed capacity of turbines where such flow may be realized for three months. Figure 8.6 shows seasonal variation of monthly reported flow from two sources (1967–1972; 1999–2003) with data sources discussed in Chap. 7 of this book. The daily average flow rate from both sources is 1560 m^3 s^{-1} which is in line with other reports.

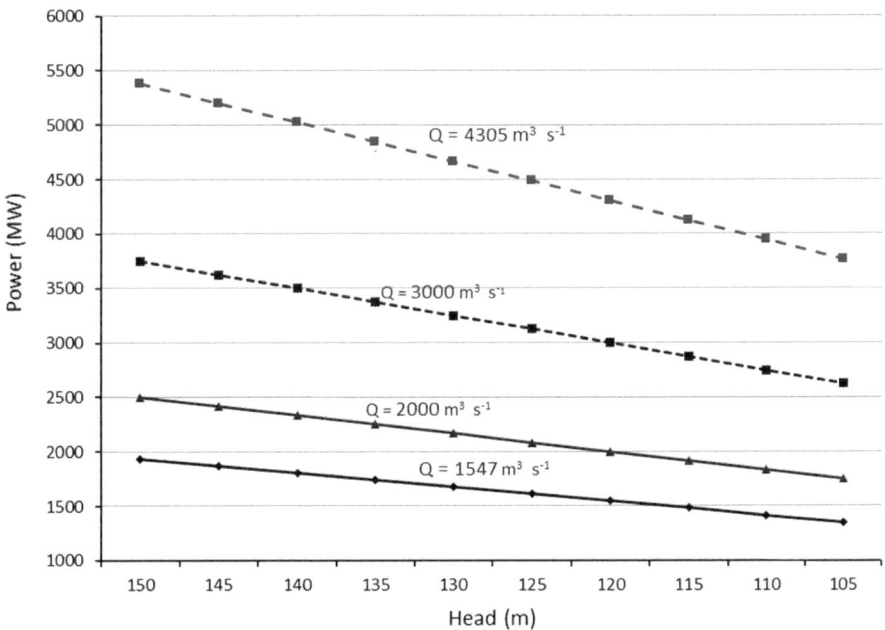

Fig. 8.5 Available hydropower for a range of GERD reservoir head and four discharge rates

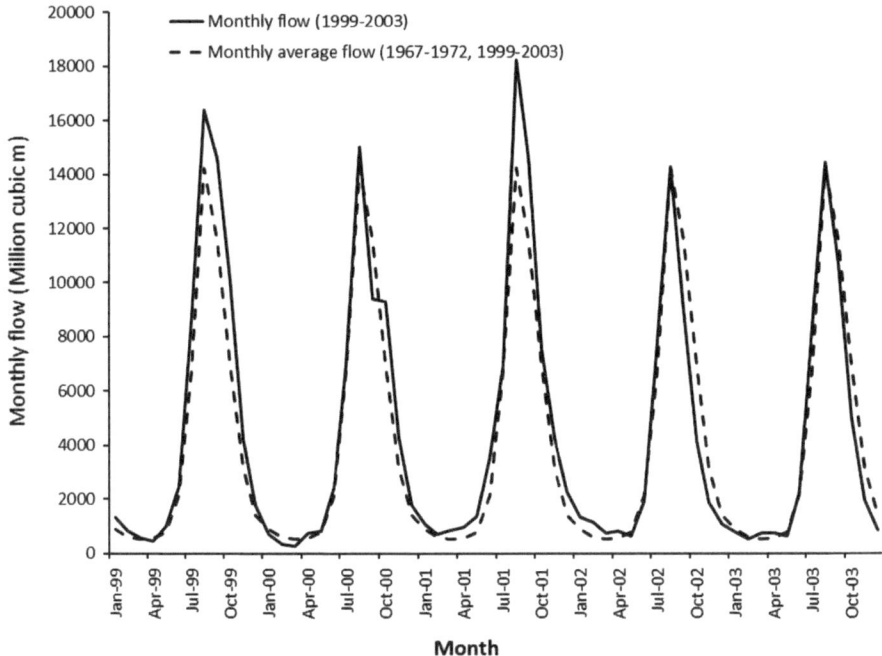

Fig. 8.6   Blue Nile River monthly flow rates at the border (GERD)

8.6   Reservoir Regulation Schedule

Reservoirs are operated with predetermined regulation or operation schedules that vary seasonally along with the hydrology of the region and defined ranked objectives. Regulation schedules have built in accommodations for dam safety, power generation, irrigation, flood control and downstream obligations. Downstream obligations are for regulatory environmental releases or satisfy downstream water rights. The importance of ecological considerations in dam operations meeting multi-objectives have been reported (Jager and Smith 2008). Short term weather and climate prediction need to be factored in the regulation for successful dam/reservoir operation including safety. Although GERD priorities or operation guidelines are not known to the authors, possible operation guidelines are:

(a) Minimum flow
(b) Preferred reservoir water level
(c) Seasonally varying minimum flow
(d) Seasonally varying preferred reservoir water level
(e) Combination of flow and water level criteria.

   Dam operation with minimum release operation schedule for power generation and downstream supply will reflect reservoir storage and water level variation with changes in annual flow volumes. Hydrologic variation and dam operation simulation of the GERD reservoir was performed with minimum discharge of 1500 and 1600 $m^3 s^{-1}$. Excess of full storage (73,000 million m^3) and stage above 645 m asl initiates additional releases to downstream and maximum power generation at the same time. Tesemma (2009) performed simulation of operation under two minimum release scenarios using 61 years of annual flow, 1912–1983 and reported Blue Nile river annual flow data from 1912 to 1996 (Fig. 8.7). This study that applied the RiverWare model to simulate flow through the reaches of the Nile has reported on hydropower and flow impact of GERD filling using multiple traces of hydrology (Tesemma et al. 2014). This study run scenarios of 5, 6, and 7 years filling period and produced impacts after 13 years with filling starting with Aswan High Dam initial stage of 173 m asl.

Operation simulation was performed using historical monthly average flow distribution but varying reported annual flow data. Monthly storage volume and reservoir water level was computed as a balance of monthly inflow, minimum release and average monthly difference between rainfall and ET and estimated seepage/leakage losses (Eq. 8.2). Figure 8.8 depicts the reservoir stage (water level) and storage volume relationship applied in the simulation with a polynomial relationship between water level and reservoir filled volume.

$$S_i = S_{i-1} + I_i - Q_m + (R - E)_i - L_i \qquad for\, S_i \leq 645\, m\, amsl \qquad (8.2)$$

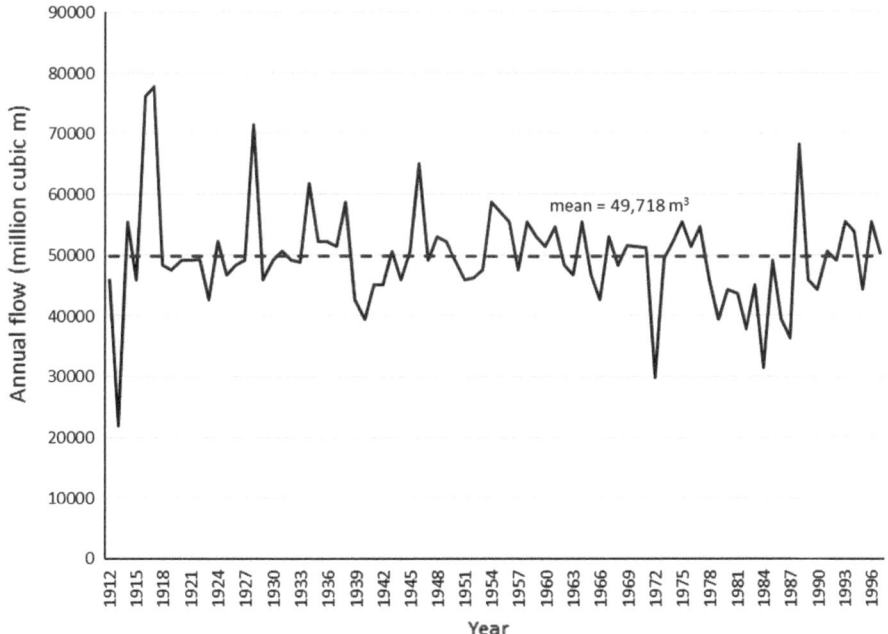

Fig. 8.7 Blue Nile River border annual flow (1912–1996; data source Tesemma 2009)

where S_i is reservoir volume at end of month i (m³), S_{i-1} is reservoir storage volume at end of month i − 1 (m³), I_i is reservoir inflow for month i (m³), Q_m is minimum discharge (m³), $(R–E)_i$ is difference in rainfall and ET for month i (m³) L is seepage/leakage losses (m³) for month i. Table 8.2 is summary of data used for simulation of reservoir operation.

Figures 8.9 and 8.10 depict continuous stage and storage monthly simulation showing loss of head during low flow periods. Continuous simulation was stopped at 49 years for 1600 m³ s⁻¹ minimum release as reservoir is drained without change in operation. During high flow periods when the reservoir is full, excess water, more than the minimum flow, has to be released. The mean annual available head for 1500 m³ s⁻¹ was 45 and 28 m for 1600 m³ s⁻¹ minimum flow operations. During high flow events, the reservoir has to release excess and generate maximum power. Since excess flow has to be released, the operation is controlled by low river flow resulting in highly fluctuating power generation.

The preliminary results of the simulation shows that power generation is dependent on minimum downstream release and the climate. Consecutive years of drought will have significant impact on power generation capacity. The impact will be extended as refilling to full reservoir level requires holding back flows.

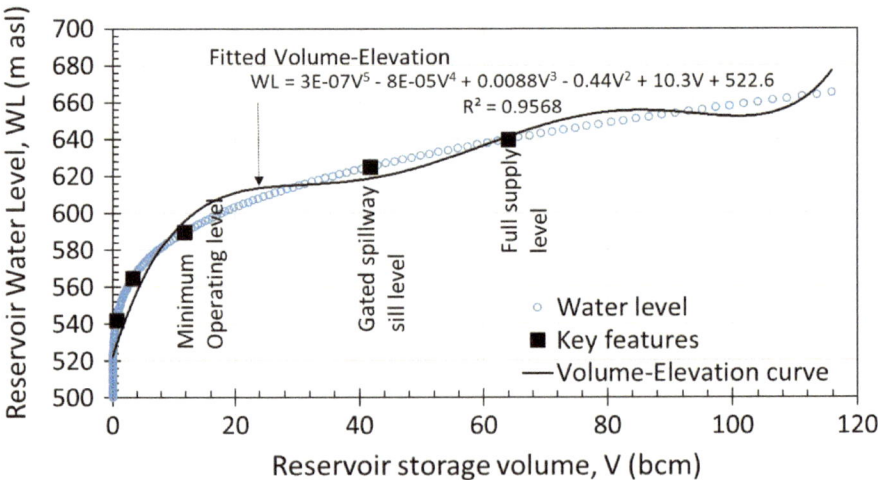

Fig. 8.8 GERD reservoir volume and water level relationship

Table 8.2 Hydrologic data used for reservoir operation simulation

Monthly mean flow	4.209 bcm
Monthly flow standard deviation	4.8 bcm
Annual mean flow	50.51 bcm
Mean annual (R–E)	−0.73 m
Mean annual seepage (L)	−0.779 m

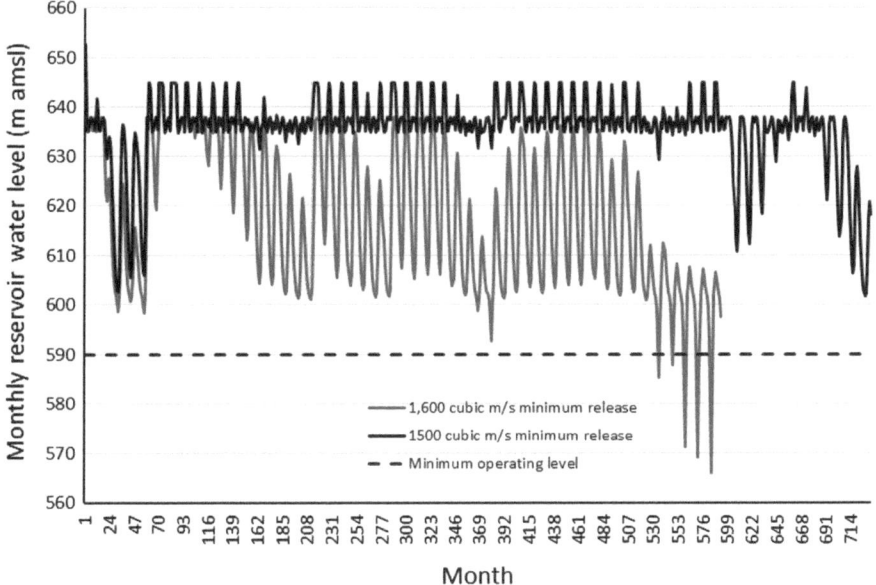

Fig. 8.9 Simulated reservoir water level for 1500 and 1600 m³ s⁻¹ minimum releases after reservoir filling

Ensembles of multiple operation scenarios can be modelled for regulation schedule alternative selection fulfilling upstream and downstream constraints, with confidence intervals.

8.7 Summary

Operation of the GERD will be challenging to market sustainable power output as a result of upstream and downstream water demand and climatic variation. Long construction period, long initial reservoir filling period, and refilling period after drought, puts challenge into the investment return of the dam. This is without considering the cost of the security of power facilities, transmission systems and the dam. Stable dam operation plan based on water rights is a necessity to meet the project's objective of power generation and marketing as downstream reservoirs. In this study, results of reservoir water level fluctuation with two minimum flow guidelines are presented with polynomial equation to calculate reservoir water level from changing volume.

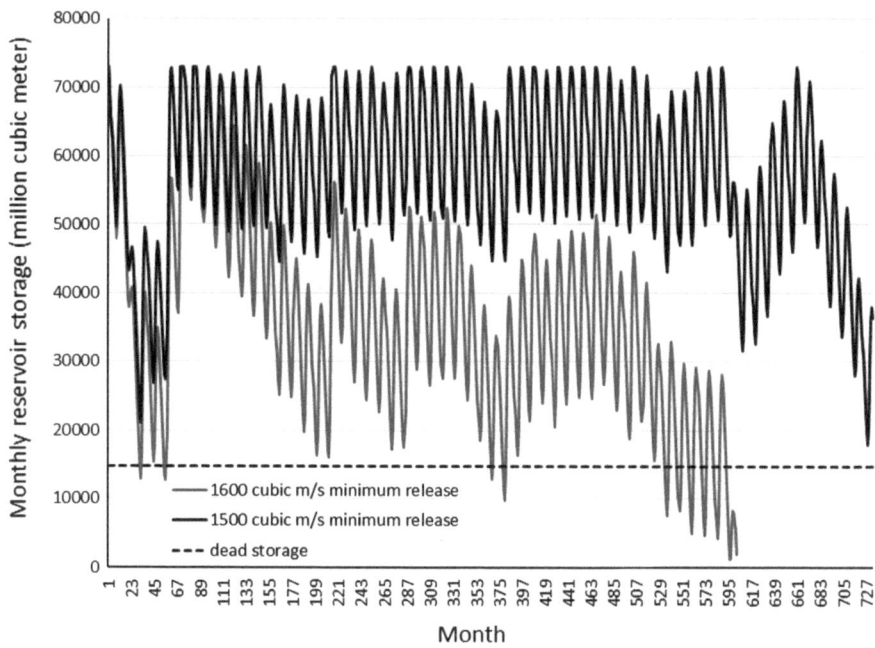

Fig. 8.10 Simulated reservoir storage for 1500 and 1600 m³ s⁻¹ minimum releases after reservoir filling

References

AbAbu-Zeid MA, El-Shibini FZ (2010) Egypt's high Aswan Dam. Int J Water Resour Dev 13:209–218 (The Blue Nile River Basin hydrology. Hydrol Process 23:3653–3660)

Abtew W, Melesse AM (2014) Ch. 33 Climatic teleconnections and water management. In: Melesse AM, Abtew W, Setegn SG. Nile River Basin ecohydrological challenges, climate and hydropolitics. Springer, New York

Abtew W, Melesse AM, Dessalegne T (2009) El Niño Southern oscillation link to the Blue Nile River basin hydrology. Hydrol Process 23:3653–3660

Block PJ, Strzepek K, Rajagopalan B (2007) Integrated management of the Blue Nile basin in Ethiopia. IFPRI Discussion Paper 00700. Colorado University, Boulder, Co

Conway D (2000) The climate and hydrology of the Upper Blue Nile River. Geogr J 166(1):49–62

Degefu GT (2003) The Nile historical legal a, Victoria, Canadand development perspectives. Trafford Publishing

Deluca J (2010) U.S.-Mexico water agreement. VOA Editorials 12 December 2010

Deneke TT (2014) Ch. 24 processes of institutional change and factors influencing collective action in local water resources governance in the Blue Nile basin of Ethiopia. In: Melesse AM, Abtew W, Setegn SG (eds). Nile River Basin ecohydrological challenges, climate and hydropolitics. Springer, New York

Eldaw AK, Salas JD, Garcia LA (2003) Long range forecasting of the Nile river flows using climate forcing. J Appl Meteorol 42(7):890–904

Eltahir EAB (1996) El Niño and the natural variability in the flow of the Nile River. Water Resour Res 32(1):131–137

Ethiopian Sugar Corporation (2015) Beles sugar development project. http://www.etsugar.gov.et/index.php/en/projects/belles-sugar-development-project. Accessed 14 May 2016, published 31 July 2015

Gissila T, Black E, Grimes DIF, Slingo JM (2004) Seasonal forecasting of the Ethiopian summer rains. Int J Climatol 24:1345–1358

International Rivers (2013) https://www.internationalrivers.org/resources/ethiopia%E2%80%99s-biggest-dam-oversized-experts-say-8082. Accessed 14 May 2017

IPoE (2013) International panel of experts on grand Ethiopian Renaissance Dam project. Final Report. Addis Ababa, Ethiopia

Jager HI, Smith BT (2008) Sustainable reservoir operation: can we generate hydropower and preserve ecosystem values. River Res Appl 24:340–352

Kem JD (2013) How deregulated markets influence hydro revenue and downstream flow. Hydro Review. www.hydroworld.com

Li X, Guo S, Liu P, Chen G (2010) Dynamic control of flood limited water level for reservoir operation by considering inflow uncertainty. J Hydrol 391(1–2):124–132

Liersch S, Koch H, Hattermann FF (2017) Management scenarios of the Grand Ethiopian Renaissance Dam and their impacts under recent and future climates. Water 9, 728 https://doi.org/10.3390/w9100728

Mulat, Moges (2014) Assessment of the impact off the Grand Ethiopian Renaissance Dam on the performance of the High Aswan Dam. J Water Resour Prot 6:583–598

Putter TD, Loutre MF, Wansard G (1998) Decadal periodicities of Nile River historical discharge (A.D. 622–1470) and climatic implications. Geophys Res Lett 25(16):3193–3196

Salini-Impregilo. Grand Ethiopian Renaissance Dam Project. http://www.salini-impregilo.com/en/projects/in-progress/dams-hydroelectric-plants-hydraulic-works/grand-ethiopian-renaissance-dam-project.html. Accessed 15 Oct 2016

Tesemma ZK (2009) Long term hydrologic trends in the Nile basin. Masters thesis, Cornell University

Tesemma ZK, Mersha A, Wheeler K (2014) Reservoir filling options assessment for the Grand Ethiopian Renaissance Dam using a probabilistic approach. Grand Ethiopian Renaissance Dam Workshop. Abdul Latif Jamal World Water and Food Security Lab, MIT, MA, 13–14 Nov 2014

Zambon RC, Barros MTL, Yeh WWG (2016) Impacts of the 2012–2015 drought on the Brazilian hydropower system. World Environ Water Resour Congr 2016:82–91

Chapter 9
Dialogue and Diplomacy Through the Construction of the Grand Ethiopian Renaissance Dam

Abstract Ethiopia unilaterally launched the construction of the Grand Ethiopian Renaissance Dam (GERD) on the main Blue Nile River (Abay) in April 2011. Since then, a series of negotiations have been conducted between Ethiopia, Egypt and Sudan at expert, ministerial and head of state levels, mainly to address Egypt's concern on potential reduction of river flow as a result of the dam. The GERD is portrayed by Egypt as threat for water security while Sudan appears comfortable with immediate access to both water and power. Sudan has tacitly approved Ethiopia's desire to build the dam anticipating benefits from a dam close to its border that could settle out sediment and control flooding. Other concerns include rate of initial dam filling, dam operations, and potential dam failure. There were four major agreements between Ethiopia, Sudan and Egypt to address these concerns: formation of International Panel of Experts, signing of Declaration of Principles by heads of states, meeting of foreign and irrigation and water ministers on implementation of IPoE recommendations, approval of water resource and hydropower simulation study by international consultants. Several meetings and consultations have been conducted without tangible outcome because of the complexity of the problem. Several see-saw of agreements and disagreements, cooperation and concern, have been reported albeit any clear future path. At this time, after reviewing the preliminary report of the study plans by the French firms, Ethiopia has pulled out of the support for the study. Significant domestic political instability in Egypt, Sudan and Ethiopia during the construction of GERD has been contributing to the slow and uncertain progress of dialogue and diplomacy.

Keywords Nile river · Blue Nile · Ethiopia · Egypt · Sudan
Grand Ethiopian Renaissance dam · Water agreements · Water conflict
Agreements on GERD · Transboundary rivers

9.1 Introduction

The Nile River is the longest river in the world, 6853 km, crossing through eleven countries from east and east-central Africa to the Mediterranean Sea. The drainage area is about 3.17 million km^2 (FAO 2007) (Fig. 9.1). The major sources are the Blue Nile (Abay) from Ethiopia and the White Nile from drainage areas in Burundi, Rwanda, Kenya, Democratic Republic of the Congo, Tanzania, Uganda, South Sudan and Sudan. In general the sources of water can be regionalized as the Ethiopian plateaus, the Equatorial Lake Region and the Bahr El Ghazal Basin in Sudan and South Sudan. Ethiopia contributes 85% of the Nile flow through the Blue Nile, Atbara and Sobat. The Nile river system can be identified in four regimes; water source, water accumulation (energy source), water losing and water consuming (Moges and Gebremichael 2014). These regimes are identifiable from source to consumption with various levels of dependence on Nile water. Estimated historical annual average flow of the Nile at Aswan is 84.1 billion m^3 (Sutcliffe and Parks 1999).

The Nile basin is shared by eleven countries with Eritrea and South Sudan being the recent new members. GERD is being built in the Blue Nile basin of Ethiopia that contributes about 60% of the Nile flow while taking only 10% of the Nile drainage area. The Blue Nile is also among the highly populated and intensively farmed regions of Ethiopia with mounting water demand for food production and power generation.

The history of the Nile water rights issue can be divided into pre-colonial, colonial and post-colonial periods. Pre-colonial period water rights can be assumed that both upstream and downstream societies did not have concern of water rights but a lot of concern to natural climatic variations as droughts and floods. Colonial period is marked by British and Belgian colonial control of the riparian countries, except Ethiopia. British controlled Egypt, 1882–1956; the Sudan, 1889–1956; Uganda, 1894–1962; Kenya, 1895–1964 and Tanzania, 1919–1961. Belgium controlled Burundi, 1916–1962; Rwanda, 1922–1961 and DRC, 1908–1960.

Colonial period water treaties and agreements did not include Ethiopia due to lack of a strong social, economic, political and military developmental stage to understand water rights and future needs and push a water right claim. The post-colonial period is characterized by progressive societal awareness and population growth, water control technology advancement and increasing upstream water needs and demands. The GERD and several other planned projects in Nile riparian countries are a major reflection of this condition in the basin. In this chapter, few of the major dealings as meetings, dialogues and diplomacy, between Ethiopia, Egypt and Sudan, that arise from the construction of the GERD are presented with reviews.

9.2 Deals and Concerns on GERD

The secrecy under which the GERD was initiated has sent anxiety to Sudan and Egypt from the beginning as they depend on the Nile waters. The lack of information on the design, objective and operation of the dam increased downstream concern.

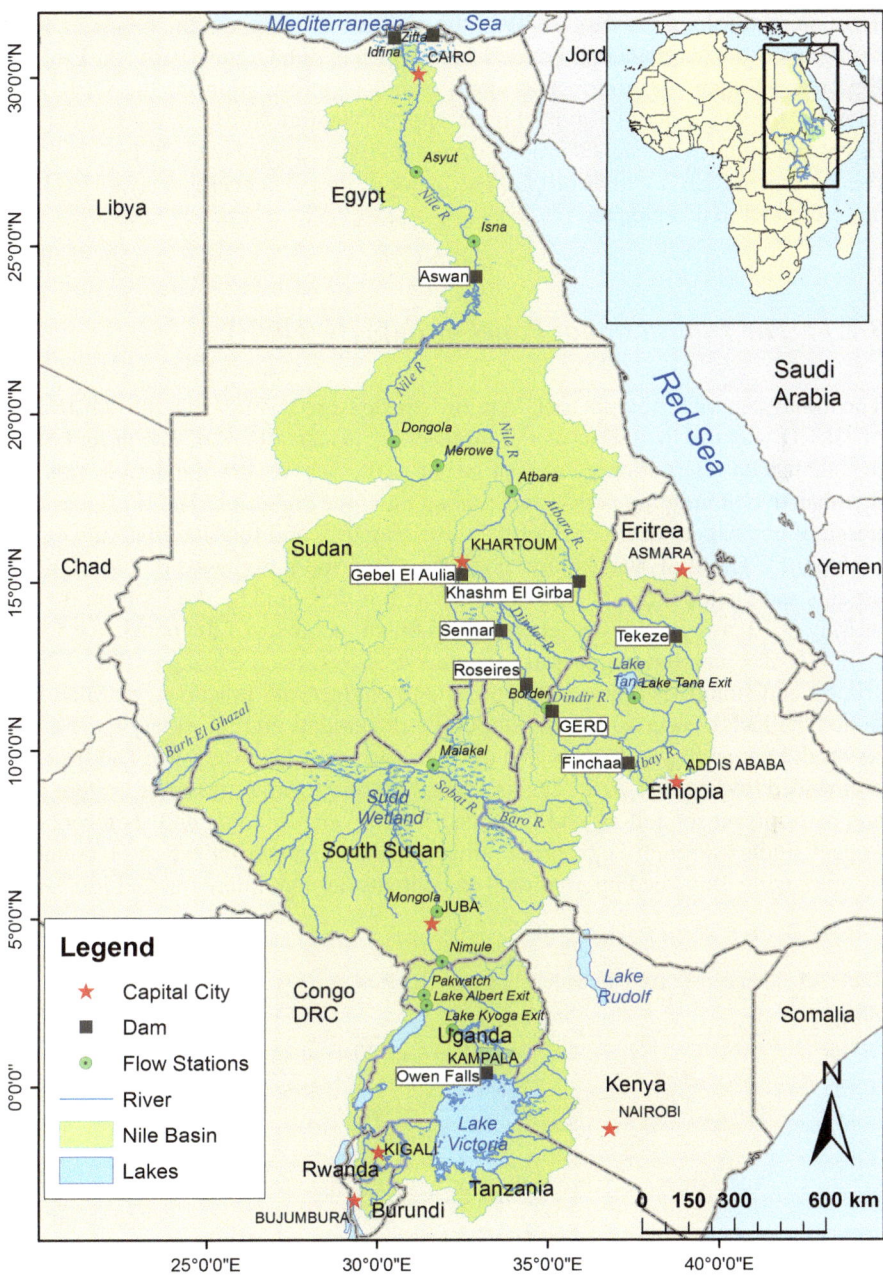

Fig. 9.1 The Nile Basin and GERD location

The potential role of the dam to affect the political structures in the region is on the background of the concerns. Several meetings have undergone between Ethiopia, Egypt and Sudan for Ethiopia to share technical information on the GERD and alleviate the concerns of Egypt. Sudan appears neutral and later the rift between Egypt and Sudan has reached critical state with Egypt asking to exclude Sudan from the dealing on GERD with Ethiopia. The major deals and concerns relevant to GERD are presented in this chapter.

9.2.1 The International Panel of Experts (IPoE)

The initial step that Ethiopia took to relieve the tension arising from the construction of GERD was the invitation of the ministers of water affairs of Egypt and Sudan to establish an International Panel of Experts (IPoE) to review the dam design with the objective of building confidence among the three countries. The IPoE was composed of two experts from each of the three countries and four international experts (Table 9.1). The IPoE was launched in May 2012 and it reviewed dam design documents and visited the dam site. Final report was delivered by the IPoE on the 30th of May 2013, as a consensus report signed by all members of the panel.

The IPoE Final Report generally concluded GERD adhered to international design criteria and standards, codes, guidelines and engineering practices; but also mentioned the lack of access to other important documents and information. The IPoE report did not address upstream and downstream impacts of the dam. The IPoE recommended two studies on the Eastern Nile System, (a) water resource system study and hydropower modelling and (b) transboundary environment and socioeconomic impact studies through appropriate arrangement as agreed by the three countries, by employing renowned international consultants through international bidding pro-

Table 9.1 Members of the International Panel of Experts

Origin	Name	Expertise
Egypt	Dr. Sherif Mohamady Elsayed	
Egypt	Dr. Khaled Ahmed	
Ethiopia	Eng. Gedion Asfaw	
Ethiopia	Dr. Yilma seleshi	
Sudan	Dr. Ahmed Eltayeb Ahmed	
Sudan	Eng. Deyab Hussien Deyab	
International	Dr. Bernard Yon	Environmental expert
International	Mr. John D. M. Roe	Socio-economics expert
International	Mr. Egon Failer	Dam engineering expert
International	Mr. Thinus Basson	Water resources and hydrological modeling expert

cess. Egypt, Ethiopia and Sudan met several times to plan the implementation of the IPoE recommendations by selecting international consultants per the agreement of the three countries through international bidding. Three meetings were held in Khartoum on November 4, 2013, December 8–9, 2013 and January 4–5, 2014.

Differences surfaced between Ethiopia and Egypt concerning the roles and the need for international experts in the implementation of the recommendations. The main differences were that Egypt wanted to introduce international experts working in parallel to the national experts and elevate them as tie breaker when differences arise. The second issue presented by Egypt related to "principles for confidence building" was declined by Ethiopia. The principles refer to issues that contradict with the Cooperative Framework Agreement Ethiopia ratified and was also signed by six other Upper Nile riparian countries on the Nile River. These and a series of disagreements between Ethiopia and Egypt with Sudan's apparent indifference are presented in a summary (Ministry of Water and Energy and Federal Democratic Republic of Ethiopia 2014). Responding to a statement by Group of the Nile Basin of Cairo University, the Ethiopian National Panel of Experts summarized general Ethiopian positions on various topics of concern to Egypt (Ministry of Water, Irrigation and Energy 2013). The following is extract on Ethiopian position on GERD related to Nile water that we have come across.

1. *"The GERD neither consumes nor diverts water to another basin"* apparently referring to Egypt's out of the Nile Basin water transfer in Toshka and Sinai. Out of basin diversion is discussed in Chap. 2 in this book.
2. *"The evaporation loss GERD incurs is significantly lower than the amount of water that the GERD saves from evaporation loss. The saving is from reduced flooding resulting in reduced seepage and evaporation."*
3. *"The GERD regulated flow brings benefits to Egypt and Sudan through flood protection, irrigation expansion, water use efficiency, sediment load reduction, affordable clean power trade, energy uplift and navigation."*
4. *"Concerning dam safety, dam design, construction and management follows international standards. Dam site is not earth quake prone and dam filling initiated earthquake is accounted in design".*
5. *"Hydrology: flow data used includes wet and dry years and so spillway design flood, Probable Maximum Flood (PMF) and diversion flood estimates have fulfilled the highest safety requirement that International Commission on Large Dams (ICOLD) recommends"*
6. *"The GERD has 74 billion cubic meter (BCM) storage capacity and about 60 BCM live storage. The 14 BCM is reserved to be filled by the sediment. The 60 BCM is mostly renewable water source that will be released every year. If GERD filling coincides with wet years, there is no concern of downstream flow reduction. If years turn out to be dry years, filling strategy will be revised. The 130 BCM storage capacity of the High Aswan Dam (HAD) in Egypt was mentioned as reserve water to overcome any downstream flow reduction due to GERD filling."*
7. *"Those people who are fear mongering with claims that so many hectares will be affected, so many farmers will be out of work are only doing great disservice*

to their own people. The facts are otherwise, as shown above. Ethiopia is a responsible nation and the design is adequate and has robust filling strategy that does not lead to any appreciable harm during filling period in the worst case combination that HAD reached minimum level and dry year occurs during filling. Again, the existing storage volume of HAD (twice the annual volume of Nile flow) has the capacity to absorb any potential multi-year shocks caused during the infilling phase of the GERD."

"Benefits to the Sudan are (a) mitigate flood risk and reduce flood damage to life and property; (b) the GERD will offer regulated flow helping Sudan conserve water and expand irrigation; (c) GERD storage will mitigate drought impacts in Sudan and Egypt; (d) GERD helps reduce water losses and cost of sediment laden canal maintenance; (e) Sudan's Roseires, Sennar and Merowe dams energy generation will be increased; (f) Sudan could generate new hydropower from run of the river system and benefit from the transmission lines for potential renewable energy generation."

"Benefits to Egypt are (a) GERD increases the total storage of Nile water benefiting Egypt during drought and flood impact and water loss during wet years; (b) GERD will benefit the Nile basin in flood routing; (c) GERD will regulate flows of the Blue Nile; (d) With GERD, evaporation from HAD will reduced to 9.5 BCM yr^{-1} from about 10.8 BCM yr^{-1}; GERD will extend the life of HAD by a century by holding back sediment; (e) GERD flow regulation will enhance navigation downstream; (f) power grid can potentially be used by renewable energy."

"Regional and global benefits are (a) GERD provides clean energy reducing CO_2 emission that would be generated for equivalent power generation from fossil fuels; (b) A number of Global Circulation Models (GCM) estimate that climate change will increase rainfall in the Ethiopian highlands where the GERD will help to manage the excess water; (c) Energy generated by the GERD will enhance regional and economic integration; (d) GERD provide power pool and flexibility; (e) GERD will increase regional trust, cooperation and development."

Despite several conflicting statements from interviews of various officials, the above statements are the only coherent statements from the Ethiopian side on GERD. Ethiopia argues GERD is solely for hydropower generation and did not regard water rights as a relevant core issue.

9.2.2 Declaration of Principles

To alleviate Egypt's concern on the GERD, the leaders of Egypt (Abdel Fattah al-Sisi), Ethiopia (Hailemariam Desalegn) and Sudan (Omar al-Bashir) signed the Declaration of Principles. Although it was heralded as a break through agreement, the core issue of water rights was skirted as has been in the other engagements that did not participate Ethiopia with the intent of downstream countries to maintain the statuesque. The most important question to be explicitly addressed was "Does Ethiopia has right to use any water from the Nile?" Both Ethiopia and the downstream countries avoid to answer this question. Review of this agreement is essential as part of Nile water rights and the GERD. The March 23, 2015 Declaration of Principles is presented as follows.

"Agreement on Declaration of Principles between Ethiopia, Egypt and Sudan

March 23, 2015, Khartoum

Agreement on Declaration of Principles between The Arab Republic of Egypt, The Federal Democratic Republic of Ethiopia And The Republic of the Sudan on the Grand Ethiopian Renaissance Dam Project (GERDP)
Preamble

Mindful of the rising demand of the Arab Republic of Egypt, the Federal Democratic Republic of Ethiopia and the Republic of Sudan on their transboundary water resource, and cognizant of the significance of the River Nile as the source of livelihood and the significant resource to the development of the people of Egypt, Ethiopia and Sudan, the three countries have committed to the following principles on the GERD:

I—Principles of Cooperation

- *To cooperate based on common understanding, mutual benefit, good faith, win-win and principles of international law.*
- *To cooperate in understanding upstream and downstream water needs in its various aspects.*

II—Principle of Development, Regional Integration and Sustainability

- *The Purpose of GERD is for power generation, to contribute to economic development, promotion of transboundary cooperation and regional integration through generation of sustainable and reliable clean energy supply.*

III—Principle Not to Cause Significant Harm

- *The Three Countries shall take all appropriate measures to prevent the causing of significant harm in utilizing the Blue/Main Nile.*
- *Where significant harm nevertheless is caused to one of the countries, the state whose use causes such harm shall, in the absence of agreement to such use, take all appropriate measures in consultations with the affected state to eliminate or mitigate such harm and, where appropriate, to discuss the question of compensation.*

IV—Principle of Equitable and Reasonable Utilization

- *The Three Countries shall utilize their shared water resources in their respective territories in an equitable and reasonable manner.*

- *In ensuring their equitable and reasonable utilization, the Three Countries will take into account all the relevant guiding factors listed below, but not limited to the following outlined:*
  a. *Geographic, hydrographic, hydrological, climatic, ecological and other factors of a natural character;*
  b. *The social and economic needs of the Basin States concerned;*
  c. *The population dependent on the water resources in each Basin State;*
  d. *The effects of the use or uses of the water resources in one Basin State on other Basin States;*
  e. *Existing and potential uses of the water resources;*
  f. *Conservation, protection, development and economy of use of the water resources and the costs of measures taken to that effect;*
  g. *The availability of alternatives, of comparable value, to a particular planned or existing use;*
  h. *The contribution of each Basin State to the waters of the Nile River system;*
  i. *The extent and proportion of the drainage area in the territory of each Basin State.*

V—Principle to cooperate on the First Filling and Operation of the Dam

- *To implement the recommendations of the International Panel of Experts (IPOE), respect the final outcomes of the Technical National Committee (TNC) Final Report on the joint studies recommended in the IPOE Final Report throughout the different phases of the project.*
- *The Three Countries, in the spirit of cooperation, will utilize the final outcomes of the joint studies, to be conducted as per the recommendations of the IPoE Report and agreed upon by the TNC, to:*
  a. *Agree on guidelines and rules on the first filling of GERD which shall cover all different scenarios, in parallel with the construction of GERD.*
  b. *Agree on guidelines and rules for the annual operation of GERD, which the owner of the dam may adjust from time to time.*
  c. *Inform the downstream countries of any unforeseen or urgent circumstances requiring adjustments in the operation of GERD.*

- *To sustain cooperation and coordination on the annual operation of GERD with downstream reservoirs, the three countries, through the line ministries responsible for water, shall set up an appropriate coordination mechanism among them.*
- *The time line for conducting the above mentioned process shall be 15 months from the inception of the two studies recommended by the IPoE.*

VI—Principle of Confidence Building

- *Priority will be given to downstream countries to purchase power generated from GERD.*

VII—Principle of Exchange of Information and Data

- *Egypt, Ethiopia, and Sudan shall provide data and information needed for the conduct of the TNC joint studies in good faith and in a timely manner.*

VIII—Principle of Dam Safety

- *The Three Countries appreciate the efforts undertaken thus far by Ethiopia in implementing the IPoE recommendations pertinent to the GERD safety.*
- *Ethiopia shall in good faith continue the full implementation of the Dam safety recommendations as per the IPoE report.*

IX—Principle of Sovereignty and Territorial Integrity

The Three Countries shall cooperate on the basis of sovereign equality, territorial integrity, mutual benefit and good faith in order to attain optimal utilization and adequate protection of the River.

X—Principle of Peaceful Settlement of Disputes

The Three Countries will settle disputes, arising out of the interpretation or implementation of this agreement, amicably through consultation or negotiation in accordance with the principle of good faith. If the Parties are unable to resolve the dispute thorough consultation or negotiation, they may jointly request for conciliation, mediation or refer the matter for the consideration of the Heads of State/Heads of Government.

This agreement on Declaration of Principles is signed in Khartoum, Sudan, on the 23 of March, 2015, by the Arab Republic of Egypt, The Federal Democratic Republic of Ethiopia and the Republic of Sudan.
For the Arab Republic of Egypt - Abdel Fattah El Sisi - President of the Republic
For the Federal Democratic Republic Ethiopia - Hailemariam Desalegn - Prime Minster of the Republic
For the Republic of Sudan - Omer Hassan Elbashir- President of the Republic".

9.2.3   Review of the Principle of Declarations

Article I is general and innocuous. Article II declares that the purpose of the GERD is power generation which could, by inference, rule out water use for irrigation from GERD reservoir by Ethiopia and assert no change in annual river flow associated to the operation of GERD.

Articles III and IV are the commonly known transboundary water use laws derived from the 1966 Helsinki International Law Association convention and the 1997 UN convention; *equitable use and not to cause significant harm.* These are regarded by experts as conflicting concepts in time of water shortage and are biased towards downstream users. For example, the concept of harm created by downstream to upstream within state and out of state is not considered. Application of such convention norms has been alleged to favour the strong.

Article V is a critical element that deals with dam filling, dam operation and implementation of the IPoE recommended technical study of hydraulic and hydrologic simulation and environmental impacts. Dam structural design, filling and operation of reservoir relation to the economics of the dam are discussed in Chaps. 7 and 8 of this book. The return on investment of the dam depends on the number of years of construction, years of dam filling, and dam operations for optimal power generation. Depending on outcome of these variables, longer construction period, longer filling period, and sub-optimal dam operation may alleviate short-term water shortage to downstream users but could render structural damage to the dam and significant economic loss due to delayed power production.

Article VI gives priority of power purchase to downstream countries which limits the marketing options of the power generator. This may be exercised as a potential deterrent to storage of water for the lack of power demand.

Article VII is on exchange of information and data for assisting the consultants' technical work recommended by the IPoE which is a good thing if sufficient quality and quantity of data is commonly shared. However, there has been lack of cooperation from Egypt to share data on water diversions and the amount of water being discharged to the Mediterranean Sea.

Article VIII is on dam safety requiring Ethiopia to implement the IPoE recommendations with respect to dam safety. Dam safety is to every ones benefit and all requirements need to be fulfilled.

Article IX is on principle of sovereignty and territorial integrity for beneficial use and protection of the river. Territorial integrity of these countries depends both on internal and external factors. There has been a consistent allegation of neighbouring countries inciting public unrest and socio-political instability on each other. The cessation of Eritrea from Ethiopia and South Sudan from Sudan are outstanding cases of the fragile environment on territorial integrity and the Nile politics has been regarded as the background driving force. Article X is on peaceful conflict resolution with respect to this declaration. Since then, Ethiopia has accused Egypt in October 2016 for helping anti-government uprisings by its peoples. Recently, Ethiopian Prime Minister Hailemariam Desalegn accused Egypt for harbouring groups targeting his country's stability (Egypt Independent, 24 December 2016). Peaceful resolution of conflicts is best option for all. Ethiopia's ethnic based structure and Article 39 of its constitution has been regarded by internal opposition as a set up for potential territorial disintegration by external adversaries. The Nile and other rivers will be source of conflict with such potential outcome that can spill over to regional conflict.

9.3 Implementation of Agreements

9.3.1 International Consultants Selection, Approval and Current Status

In April 2015, the water ministers of Ethiopia, Egypt and Sudan selected two international consultants to address the two IPoE recommended studies; water resource system/hydropower model simulation and transboundary environment and socioeconomic impact studies of the Easter Nile system. The Dutch consulting firm Deltares and the French consultancy Firm BRL were selected with 30–70% share of the study. After delays of the start of the study and wrangling between Ethiopia and Egypt, Deltares withdrew in September, 2015 citing insufficient guarantee to do independent and quality study (ahramOnline, 10 September 2015).

Foreign affairs and irrigation/water ministers from Egypt, Ethiopia and Sudan met in Khartoum, Sudan, from December 27 to 29, 2015 to plan the implementation of the Declaration of Principles signed earlier in March and reselect the international consultants. Several issues were raised by Egypt without reaching consensus. The meeting agreed on the selection of French firms BRL and Artelia to do the studies. Studies were planned to take 15 months and start in February 2016 after a triparty meeting but no action was taken to initiate the work.

After the December 29, 2015 agreement, Egypt and Ethiopia expressed different understandings of the meeting. Ethiopia complained that Egyptian delegation went on media campaign and distortion of information right after the meeting, according to the Ministry of Water and Energy and Federal Democratic Republic of Ethiopia (2014). Egypt's expressed interests but not accepted by Ethiopia were (1) signing legally binding guarantee of Egyptian water right, and (2) Egyptian participation of dam administration. If consensus is not achieved Egypt has threatened to take the issue to the UN Security Council as a threat to national security (https://www.stratfor.com/analysis/ethiopia-makes-progress-nile-dam-project accessed 4 January 2016). Egypt has sought change in dam design to increase the number of outlets to guaranty discharge capacity where Ethiopia rejected the proposal according to the state-run Ethiopian Broadcasting Corporation.

In January 2016, the French consultancy firms BRL and Artelia offered proposal of the study to the three countries. After several meetings and delays; Ethiopia, Sudan and Egypt signed an agreement in October 2016 approving the two firms to perform a water resources study and hydropower simulation modelling and downstream environmental and social impact of the dam (Water Power and Dam Construction, 27 October 2016). The study was estimated to cost $4.9 million US dollars and scheduled to start in November 2016. The study was expected to take 11 months under the supervision of the Tripartite National Committee (TNC). U.K based law firm, Corbett & Co was approved to manage legal matters for the tripartite technical committee. But, by October 2016, Ethiopia accused Egypt for fermenting past and current divisive public uprising in Ethiopia raising the level of hostility shading uncertainty on future dialogue and diplomatic engagements.

9.4   Recent Media Coverage on GERD, Ethiopia, Egypt and Sudan

9.4.1   Coverage on Dialogue and Diplomacy

Local and international media outlets have been covering the status of regional diplomacy and the internal political implications of GERD. Some of the media coverages has been inflammatory, triggering swift and often undesirable response towards future co-operation. By December 25, 2016 Egyptian deputy foreign minister, Ibrahim Yusri, presented legal argument to the Egyptian parliament to nullify agreement on GERD with Ethiopia and Sudan as it contradicts Egypt's constitution and interests. At the same time, Ethiopian foreign minister, Workineh Gebeyehu, requested response from Egypt about ending the support and harbouring of Ethiopian opposition activity in Egypt (ahramonline 27 December 2016). The escalation of political rhetoric and alleged accusations on meddling into internal affairs cast shadow on the progress of dialogue and negotiations while creating grounds for potential conflict. In January 2017, a group of Egyptian lawyers and political activists are reported planning to challenge their president's agreement on the March 2015 Declaration of Principles on GERD. Their concern is that the agreement counters Egypt's interests in water and electricity security (ESAT TV news 19 January 2017).

Egyptian farmers' reportedly expressed fear of water shortage and the government tries to calm them down with a statement that the GERD has not started filling. Reports of Egyptian farmers complaints about persistent drought and water shortage and fear has been reported even though the GERD has not yet been completed (Climatechangenews.com, Accessed 18 July 2017). Without defined upstream and downstream water rights and agreements to manage and share the resource, water conflict may be inevitable driven by social pressure and political desire to take advantage of public hysteria and fear. GERD has also been used as a ripe and exploitable piece for internal political gain and rally.

On July 30, 2017 Egyptian irrigation minister gave statement under the topic Egypt is ready to resolve tensions with Ethiopia. According to him, the three countries have received partial report from the consultants (French firms Atelia and BRL) and reached consensus. Another part of the report has created disagreement related to the methodology used in the study (Ezega.com, 30 July 2017).

Recent developments show Sudan started softening its position on GERD as expressed by its foreign minister press conference speech on the event of the Ethiopian Prime minister state visit in Khartoum. He stressed the critical need of Egyptian participation in Nile water related issues (Sudan Tribune, 16 August 2017). Sudanese President Omar al-Bashir, at a joint press conference with Prime Minister of Ethiopia, Halemariam Desalegn, expressed his commitment to the 1959 Nile water sharing agreement where all Nile water is shared by Egypt and Sudan (Xinhauanet 18 August 2017).

On August 8, 2017, Ethiopia's information and technology minister stated that 60% of the dam work is completed and preparation is being made for initial power generation (XINHUANET 8 August 2017). On September 25, 2017 Middle East Monitor reported that Egyptian government sources reported failure of technical talks on the Grand Ethiopian Renaissance Dam and Egypt is studying political and diplomatic moves to maintain its water right. Sticky points are mentioned to be dam storages along the Nile and water withdrawal along the river system.

On October 18, 2017 ministers of water and irrigation of Ethiopia, Sudan and Egypt met in Addis Ababa to discuss the slow progress of the two commissioned technical studies by French firms Atelia and BRL after visiting the dam site the previous day. In attendance were Ethiopian minister of water, irrigation and electricity, Sileshi Bekele; Sudanese minister of water resources, irrigation and electricity affairs, Muataz Musa; and Egyptian minister of irrigation, Mohamed Abdel-Aty. The delay of technical study, dam size, dam filling and dam operations were some the topics of discussion (XINHUANET 19 October 2017).

A two-day meeting of the tripartite technical committee of Ethiopia, Egypt and Sudan met in Cairo on November 11 and 12, 2017 to discuss introduction material produced by the study firms, Artelia and BRL. The meeting ended in disagreement with Egypt approving the introduction report and Ethiopia and Sudan's irrigation ministers' asking for amendments that would change the report. Egypt raised its concern in the delay of the technical study and maintaining its current water share of 55.5 billion cubic meter (XINHUANET 13 November 2017). At the same meeting, Sudan's representative, minister for water resources, irrigation and electricity, questioned the credibility of the study because it did not account full share of Sudan. According to the minister, the 1959 treaty allows 55.5 to Egypt and 18.5 BCM to Sudan and stated Egypt owes Sudan 6.5 BCM from a separate agreement. The minister stated that the main concern of Egypt is that it will lose Sudan's share as a result of GERD (Sudan Tribune 21 November 2017). In response, the Egyptian foreign minister stated that no Sudanese water share comes to Egypt except excess during flooding which Egypt has to release to relive the Aswan Dam.

The Egyptian foreign minister informed his USA counterpart that the GERD technical issue is at a deadlock and both confirmed their commitment to the Declaration of Principles between Ethiopia, Sudan and Egypt signed in 2015, (Ahramonline 23 November 2017). The Washington Post (23 November 2017) reported that tensions are flaring up between Egypt, Ethiopia and Sudan citing that Egyptian media is labelling Sudan taking Ethiopia' side because of its border issue with Egypt. In a meeting of the Ethiopian and Egyptian foreign ministers in Addis Ababa in the last week of December 2017, Egypt proposed the introduction of the World Bank in the technical negotiations on GERD. There is precedence where the World Bank was part of a transboundary river water disagreements, on the Indus River between India and Pakistan (Ahramonline 04 January 2018).

It should be noted that the rest of the Nile basin countries were not part of the 1959 treaty and no water is allocated for them. The history of Nile water treaties and colonialism is a subject by itself (Abtew and Melesse 2014; Degefu 2003) and addressed in other chapters in this book.

9.4.2   Coverage on Alleged Dam Security

The Ethiopian government has stated at least twice that it caught individuals who attempted to sabotage the dam. Eritrean president gave disclaimer that reports that Eritrea is colluding with Egypt to sabotage the GERD was not factual (Middleeast-monitor 25 September 2017). The same issue stated that Sudan and Ethiopia are on alert for Egyptian strike. As long as the tension persists, the potential for water conflict is real. Ethnic conflicts within Ethiopia and opposition to the government can raise security issues for dam construction and operation. As we develop this chapter political events in Ethiopia are quickly changing; the prime minister resigned, state of emergency is reinstated; uprising continues in the various ethnic regions.

Relationships developing between Ethiopian and Qatar has led to bilateral visits and reported accounts suggest Ethiopia may look for fund from Qatar to finish the dam (africanews.com 18 November 2017). The intricate politics of the Middle East could creep into the GERD raising the stake for all interested. On November 18, 2017, Egyptian president for the second time in a month delivered a stern warning to Ethiopia, Nile dam dispute is life or death (The Washington Post 18 November 2017). On January 2, 2008 Aljazeera reported that the Egyptian president made a request to the Ethiopian prime minister to exclude Sudan from GERD related negotiations (http://www.aljazeera.com/news/2018/01/egypt-sudan-contentious-dam-talks-180102123313038.html). Based on the stalemate of technical study on GERD impact, Egypt's Minister of Water resources and Irrigation stated that there are many alternatives ways that cannot be declared to negotiate with Ethiopia and Sudan on the dam. He also said the dam is damaging to Egypt and is currently working to lessen the damage (Arab news 7 December 2017). The Arab League stated its extreme concern on GERD (Durame.com Accesses 29 November 2017).

The Ethiopian prime minister and delegation met their counterparts in Cairo on January 18 and 19, 2018. Egypt expressed its concern on the dam and lack of progress in the agreed upon technical study. It also suggested the introduction of the World Bank into the negotiations. Ethiopia suggested the formation of a new technical team (www.news24.com 20 January 2018). Ethiopia and Sudan are deploying their armies jointly near the GERD (ESAT 19 January 2018).

Kameri-Mbote (2007) in her study of conflict on the Nile, stated that Egypt's recent leaders have predicted their next war will be on water. But she suggested war on water can be averted through cooperation of Nile basin countries in search of new legal framework for the management of the Nile using the Nile Basin Initiative as a catalyst. The power asymmetry in the eastern Nile basin and the difference between Egyptians on how to deal with GERD and the Nile Basin Initiative, will make it hard to reach a deal on benefit sharing (Tawfik 2016).

Changes are occurring fast… Ethiopia is occupied with uprisings and emergency law with the prime minister resigning. The Sudanese president met Egypt's counterpart met on March 19, 2018 in Cairo and came into agreement to cooperate across various fields (Egypt Today 20 March 2018).

9.5 Summary

Cooperation on transboundary water resource development is primarily revolves around water rights issues where complex cases like the Nile makes piece meal agreements tricky with low chance of lasting accord. Over a century, unilateral projects on the Nile have been initiated and implemented mostly by downstream countries or their past colonial rulers. However, the perceived water right of Egypt and Sudan have been challenged by socio-political and economic changes in the upstream countries. The water right challenge is also tangled with a growing interest of upstream countries to attract international commercial agriculture that may result in increased consumptive use. In the absence of comprehensive agreement among riparian countries on basin wide water sharing and management that reflects current realities, political tension will continue to build to the brink of water conflict.

Since the ground breaking of GERD, four significant diplomatic events took place towards improving dialogue and co-operation. The first undertaking was the formation of the International Panel of Experts (IPoE) in May 2012 to review the dam design with production of a report on May 30, 2013. The March 23, 2015 Declaration of Principles agreement between the three heads of states was the second significant event followed by the December 27–29, 2015 meeting between respective foreign and irrigation and water ministers on the implementation of the Declaration of Principles. Finally the October 27, 2016 signing of the water ministers of the three countries launching a water resources study and hydropower simulation assessment of the Eastern Nile by two French engineering consultants, BRL and Artelia, was the last major event followed by a series of meetings on its specifics and execution. U.K based law firm, Corbett & Co was to manage legal matters for the tripartite technical committee. The study is to last 11 months and is to be supervised by the Tripartite National Committee of the three countries formed earlier. But, by October 2016, Ethiopia has accused Egypt for fermenting past and current divisive public uprising in Ethiopia raising the level of hostility shading uncertainty on future dialogue and diplomatic engagements.

The slow and daunting cooperative dialogue and diplomacy did not keep up with exploding population and economic growth as well as internal political pressures forcing upstream countries to commission mega-projects on the Nile and its tributaries. At the centre of the shifting landscape on the Nile water right issue is the construction of GERD in Ethiopia that may reset the hydro-politics and regional power structure. Concerning the GERD, the coming report by the two international consultants is likely to intensify disagreements between Ethiopia and Egypt. Despite being portrayed as a threat by Egyptians, GERD may have presented an opportunity to engage on constructive dialogue towards sustainable sharing of the Nile waters and their economic benefits. However, diplomatic fallout and failure to recognize the current reality may spiral to undesirable water conflict. Increasing water demand, limited resource, food insecurity and instability in the region decrease the chance for orderly resource sharing in the Nile basin. As time passes, the insufficiency of the Nile and tributary waters for the growing basin population will be realized.

References

Abtew W, Melesse AM (2014) Ch. 28 Transboundary rives and the Nile. In: Melesse AM, Abtew W, Setegn SG (eds). Nile River Basin ecohydrological challenges, climate and hydropolitics. Springer, New York

Degefu GT (2003) The Nile historical legal and developmental perspectives. A warning for the twenty-first century. Trafford, Victoria, Canada

FAO (2007) http://www.fao.org/nr/water/faonile/products/Docs/Poster_Maps/BASINANDSUBBASIN.pdf. Accessed 2 Jan 2016

Kameri-Mbote P (2007) Water, conflict, and cooperation: lessons from the Nile basin. Woodrow Wilson International Center for Scholars. No. 4

Ministry of Water and Energy, Federal Democratic Republic of Ethiopia (2014) Settling the record straight. http://www.mowr.gov.et/index.php?pagenum=0.1&ContentID=108. Accessed 5 Nov 2016

Ministry of Water, Irrigation and Energy (2013) Unwarranted anxiety The Grand Ethiopian Renaissance Dam (GERD) and some Egyptian experts hyperbole, 24 June 2013. http://www.mowr.gov.et/index.php?pagenum=0.1&ContentID=88. Accessed 4 Dec 2016

Moges SA, Gebremichael M (2014) Chapter 18 Climate change impacts and development-based adaptation to the Nile River basin. In: Melesse AM, Abtew W, Setegn (eds) Nile River basin ecohydrological challenges, climate change and hydropolitics. Springer, New York

Sutcliffe JV, Parks YP (1999) The hydrology of the Nile, IAHS, Special Publication No. 5. IAHS Press, UK

Tawfik R (2016) The Grand Ethiopian Renaissance Dam: a benefit-sharing project in the Eastern Nile? Water Int 41(4):574–592

Chapter 10
Aquatic Weeds and the Grand Ethiopian Renaissance Dam

Abstract Aquatic vegetation are constant problems on reservoirs, lakes, wetlands and waterways around the world with water hyacinth being the most problematic. In recent years water hyacinth has expanded in Lake Tana, the source of the Blue Nile River. The Grand Ethiopian Renaissance Dam (GERD) reservoir downstream will have to deal with the challenges of such aquatic vegetation. The Lake Tana water hyacinth experience and current efforts to control are discussed. Water hyacinth control methods applied in many parts of the world, as biological, chemical, mechanical and manual methods are reviewed and their potential applications to the GERD Reservoir are discussed.

Keywords Nile river · Blue Nile · Ethiopia · Egypt · Sudan · Grand Ethiopian Renaissance Dam · Aquatic weed · Water hyacinth · Reservoir · Lake Tana

10.1 Introduction

Water hyacinth (Eichhornia crassipes) is an aggressive aquatic weed with an extremely rapid reproduction rate resulting in chocking of water bodies (Fig. 10.1). It is the most problematic aquatic weed with the ability to double in mass in a few days and migrate easily because of its free floating vegetative form (Williams 2005). It is a floating plant that forms large mat, tussocks or floating islands in clumps capable of moving around with flow or wind driven disturbances. It affects the aquatic ecosystem by blocking sunlight and reducing aeration. Dense shading in water bodies and wetlands reduces light penetration depth affecting submerged vegetation. Water hyacinth impacts navigation, power generation, irrigation, transportation and canal conveyance. Its origin is believed to be South America, perhaps the Amazon basin (Barrett and Forno 1982). The decorative look made Water hyacinth attractive to be moved and introduced outside its native area. It was first introduced to the United State in late 1800 and moved on to other countries (Gopal 1987).

According to Prof. Mark Mwandosya (Nile Basin Discourse 2017), water hyacinth was introduced to Egypt in the 1890s and covered the Nile basin by the 1990s. The

© Springer International Publishing AG, part of Springer Nature 2019   147
W. Abtew and S. B. Dessu, *The Grand Ethiopian Renaissance Dam
on the Blue Nile*, Springer Geography, https://doi.org/10.1007/978-3-319-97094-3_10

Fig. 10.1 Water hyacinth in Lake Tana

weed was introduced to Lake Victoria in the Nile basin from Rwanda through the Kagera River by human activities. Despite the disruption in fishing activities and being breeding media for human pathogens, locals use water hyacinth for making furniture, paper and artefacts (Opande et al. 2004). Fishers in Lake Victoria have reported some benefit from the weed with anecdotal increase in lost fish species that started using it as breeding sanctuary. But the list of harms associated with the weed include loss of biodiversity, prevention of water oxygenation, water quality reduction, affect to tourism and recreation, increase in pest and water born vectors causing malaria, typhoid, dysentery, schistosomiasis, rift valley fever, *E. Coli* infection, bilharzia and cholera (Nile Basin Discourse 2017).

Half of China is affected by the weed with multi-sector impact; decline in native diversity, spread of human disease, and economical loss by impeding flow, paralysing navigation and damaging irrigation and hydropower facilities (Wu et al. 2007).

10.2   Water Hyacinth in Lake Tana

Lake Tana, is the beginning of the main Blue Nile River in Ethiopia. The Lake water supply, fishing, hydropower, recreational, and tourism especially with island monasteries and church with critical historical value. By 2011 about 20,000 ha were covered by water hyacinth growing close to 40,000 ha by 2014 (Kibret 2017). The weed has been expanding in the lake causing multitude of problems. Manual removal to control the progress of the weed has been reported unsuccessful.

The purpose of the GERD is stated as Hydroelectric power generation to boost Ethiopia's domestic power supply and power export to neighbouring countries. Fishery is expected to be a major industry once the GERD reservoir is filled. The project site is located about 500 km northwest of the capital Addis Ababa and 20 km from the Sudan border. The dam site is further downstream of the Blue Nile gorge in the lowland requiring large reservoir area of more than 1700 km^2. It will form a lake size of 65% of Lake Tana (Fig. 10.2). Lake Tana is currently infested with large scale water hyacinth (Fig. 10.1). A 2017 estimate of water hyacinth coverage is over 5000 ha (Global Coalition for Lake Tana Restoration, personal communication). Current control effort is manually pushing vegetation mass to the edge of the lake and also bagging and removing by volunteers. The GERD reservoir will likely be affected by aquatic weed where management plan with control method is essential. Terrestrial plants on the edges of the reservoir will emerge and the range will be determined by hydroperiod that depends on temporal reservoir water level fluctuation (Fig. 10.3). In the littoral zone, machrophytes (emergent plants), floating and submerged plants will populate with floating tussocks moving along the flow towards hydraulic structures of the dam. Water hyacinth will easily be transported from Lake Tana to the GERD. Weed control and management in Lake Tana will have a direct implication on sustainability of any similar effort in the GERD reservoir.

10.3   Water Hyacinth Controls

Water hyacinth and other floating vegetation require management to fully utilize water bodies. Best management of such invasion is to take action on the first sighting before expansion. Water hyacinth is controlled with physical or manual, biological, chemical and mechanical methods. Water hyacinth is the most important weed in South Africa. Mechanical and chemical control has been unsatisfactory with biological control being evaluated for the long term (Cilliers 1991). Weed control methods employed in different parts of the world are reviewed. Lessons drawn from aquatic weed control in subtropical climate of South Florida, USA are also discussed below as an example on the alternative management practices for aquatic weed control in the GERD reservoir and Ethiopian Lakes. Florida lakes and water bodies are commonly plagued with native and exotic floating, submerged and terrestrial vegetation with choking effects.

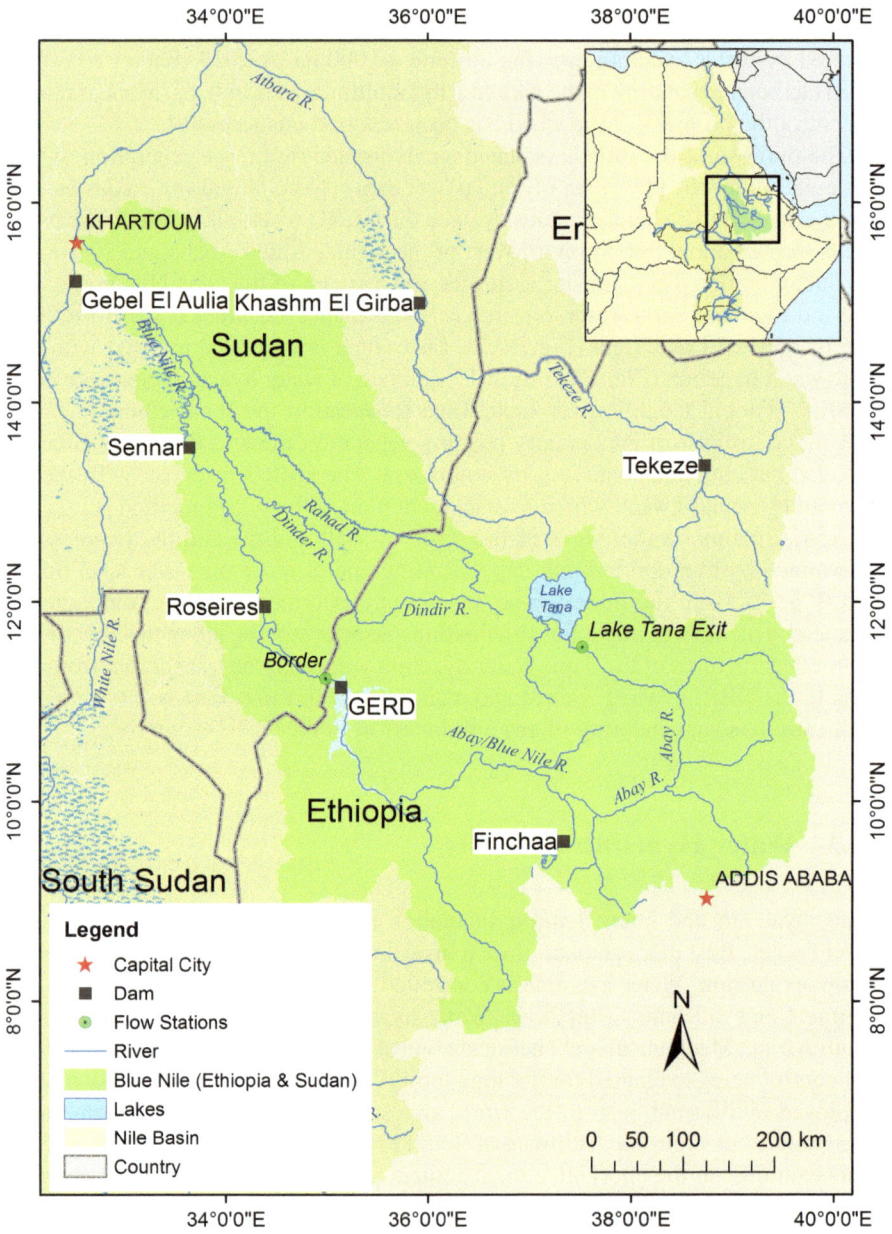

Fig. 10.2 Lake Tana and Grand Ethiopian Renaissance Dam location along the Blue Nile River

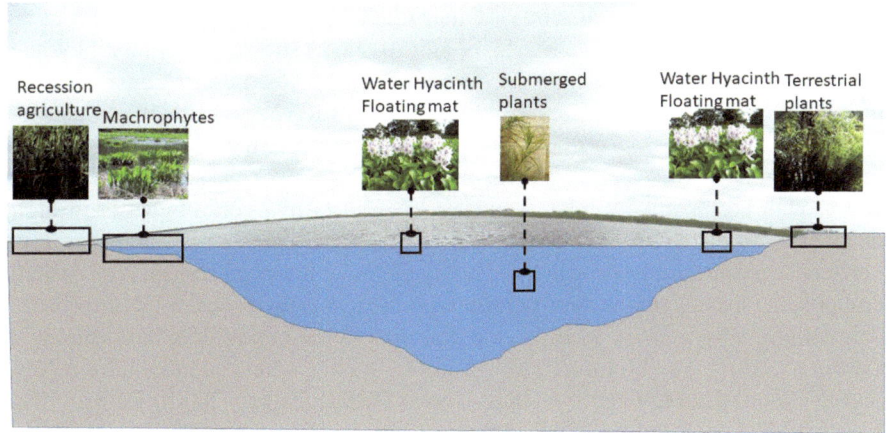

Fig. 10.3 Vegetation in and around reservoirs and lakes

10.3.1 Manual Methods

Physical or manual control of water hyacinth is the most widely used method in developing countries. According to Julien et al. (1996), manual control is good for areas less than a hectare and where labour is cheap but not viable for large scale infestation. It could be effectively applied in controlling early invasion or resurgence of aquatic weed and extensive study on how to improve effectiveness of manual removal is essential.

10.3.2 Biological Methods

Biological control is the introduction of species that feed on aquatic weed limiting growth and reproduction and cause mortality. In some instances, the biological agent is imported, reproduced and released. Success is dependent on the mass of the release. In Florida in the 1970s, two weevil specious, *Neochetina bruchi* and *Neochetina eichhorniae*, for water hyacinth and a weevil (*Neochetina affinis*) and a moth (Namangama *pectinicornis*) for water lettuce control has been introduced. Results are not very effective with the weeds continuing to be problematic and requiring additional chemical and mechanical methods (South Florida Water Management District 2008). Method of rearing and distribution of the two weevil species has been reported (Julien et al. 1999). Cercospora *rodmanii* is a fungal pathogen native to Florida that induces leaf spots, leaf necrosis and root rot on water hyacinth (Charudattan 1986).

A study of the economic value of biological control was reported in southern Benin where two species of weevil and a species of moth was applied. A 2 million US dollars cost of biological control produced a 30 million US dollar increase in income for the population (Groote et al. 2003). Other methods include using herbaceous fish and arthropods (Gopal 1987). Marlin et al. (2013a, b) showed combined biological control was better than a single species application after observations of damage to leaves by mites (*Orthogalumna terebrantis*), mirid (*Eccritotarsus catarinensis*) and weevils (*Neochetina eichhorniae*). Heard and Winterton (2000) studied water nutrient status, water hyacinth growth and weevil herbivory and showed that nutrient and growth are positively related and *Neochetina bruchi* controls better in higher nutrient concentrations and *Neochetina eichhorniae* does better control in lower nutrient concentration. According to Harley (1999), biological control is an essential component of managing water hyacinth with the two weevil species cited earlier and the moth (*S. albiguttalis*). In China the dominant weed control method is biological (Wu et al. 2007). By the year 2000, seven of the Nile basin countries have released biological control on water hyacinth; Congo (1999), Egypt (2000), Kenya (1995), Rwanda (2000), Sudan (1979), Tanzania (1995) and Uganda (1979) as reported by Julien (2000).

The process of introducing biological control into an environment requires controlled study of survival of the biological element in generations in the new environment, its impact on non-target native and economic plants and the reduction of biomass and number of leaves of water hyacinth when compared to the control. In the Invasive Plant Research Laboratory of United States Department of Agriculture (USDA), Agricultural Research Service (USDA-ARS) in Davie, Florida; biocontrol study was done on a planthopper, *Megamelus scutellaris,* which feeds on water hyacinth., The planthopper was brought from South America released in south Florida as a bio-control (Rodgers et al. 2017). The experimental study showed the planthopper reduced biomass by 67% and number of leaves by 73% compared to the control (Tipping et al. 2011). A controlled experiment in water hyacinth bio-control of demonstrated damage to plants by water hyacinth weevil (*Neochetina affinis)* and water hyacinth planthopper Megamelus *scutellaris* Fig. 10.4a–d.

Figure 10.4 Controlled bio-control experiment by USDA-ARS (a) control water hyacinth, (b) herbivory by (c) the water hyacinth weevil (*Neochetina affinis)* and (d) water hyacinth planthopper (Megamelus scutellaris) (Photo by USDA-ARS; Rodgers (2018).

10.3.3   Mechanical Methods

Mechanical removal mechanisms include removal of vegetation to the side of the water body or in a designated area in the lake. Mechanically chopping the vegetation and sinking it to the bottom is another method after herbicide application. Decomposition of the organic matter may deplete oxygen and result in fish kill. A minimum of 2 mg L^{-1} dissolved oxygen is required for fish sustenance. Mechanical systems require the initial cost of acquiring the machines and the technical capability to properly operate and maintain the system. Figure 10.5 depicts mechanical removal of floating vegetation in Florida.

Fig. 10.4 Controlled bio-control experiment by USDA-ARS **a** control water hyacinth, **b** herbivory by **c** the water hyacinth weevil (*Neochetina affinis*) and **d** water hyacinth planthopper (Megamelus scutellaris) (Photo by USDA-ARS; Rodgers (2018)

The cost of mechanical control is high. In Florida, a contracted floating harvester could cost as high as 1250 US dollars per ha (Fig. 10.6). The harvester removes floating or submerged vegetation to shore (Fig. 10.7). Removal of vegetation from shore to a disposal site is additional cost (Fig. 10.8). Mechanical control could be cost effective for relatively smaller area to control re-emergence or a new establishment of aquatic vegetation.

10.3.4 Chemical Methods

Herbicide chemicals are used to control water hyacinth in water bodies such as reservoirs, lakes, wetlands and waterways. As may be expected, there are various concerns on application safety and environmental side effects of these chemicals. Herbicides are applied on aquatic weed leaves manually, by boat or areal application. Guadalupe reservoir/dam in Mexico was 95% covered with water hyacinth. Chemical application of herbicides diquat and 2-4-D amine fully controlled the weed but two side effects were observed: (1) toxic effect on the growth of phytoplanktonic species such as zooplankton; and (2) oxygen depletion of decomposing water hyacinth (Lugo et al. 1998). In South Florida, multiple agencies conduct nuisance vegetation control

Fig. 10.5 Mechanical removal of floating vegetation in Tsala Apopka Lake, Citrus County, north central Florida (https://plants.ifas.ufl.edu/manage/why-manage-plants/tussocks-and-floating-islands/)

in lakes, wetland and waterways. Lake Okeechobee in South Florida is about the size of GERD (1780 km^2) and suffers from terrestrial, emergent and floating nuisance vegetation where continuous control is needed mostly with chemical treatment. From May 2015 to April 2016, 5668 ha of water hyacinth and water lettuce; and 3440 ha of cattails was chemically treated (Fig. 10.9). Other nuisance vegetation include torpedo grass, Phragmites and Luziola (Sharfstein and Zhang 2017). Herbicides used to control floating and submerged vegetation in South Florida area 2,4-D, diquat, fluridone, endothall and triclopyr (Rodgers et al. 2017). Application rate of herbicide on water hyacinth and water lettuce in a number of constructed wetlands in south Florida was an average of 3 L ha^{-1}. The herbicide was Reward® with diquat as active ingredient (Johnson 2008).

A combination of several methods may need to be employed for effective control depending on the type of weed and extent of coverage. Mechanical systems are good for new invasion or resurgent infestation of manageable area. Biological controls are costly at the early stages of developing and releasing the agents but mostly are cost effective in the long run. Chemical method with the appropriate regulatory control is probably the indispensable method for effective control of aquatic weeds with aggressive reproduction or growth rate. Van Wyk and van Wilgen (2002) studied cost of various methods for water hyacinth control in South Africa compar-

Fig. 10.6 Floating aquatic vegetation harvester (Photo from South Florida Water Management District)

ing biological, chemical and integrated methods and found that biological method is the most cost effective in the long term. Chemical (herbicidal) methods are the most costly and have problems of inappropriate applications and management and institutional inconsistencies. The study recommends integrated control where each method is applied where most effective and cost efficient. A study of a combined use of biological control and different rates of herbicide 2,4-D applications showed that reduced herbicide application combined with biological control provided comparative result to operational rate of herbicide application combined with biological control (Tipping et al. 2017).

10.4   Utilization of Water Hyacinth

The global expansion of the weed has initiated research in control and utilization. Areas of utilization of water hyacinth include compost, biogas and wastewater treatment in decentralized systems (Malik 2007). Bioconversion of water hyacinth to ethanol has shown some success (Nigam 2002). Other results show ethanol production from water hyacinth comparable to agricultural biomass producing 0.15 g of ethanol per gram dry weight of water hyacinth (Mishima et al. 2008). Singhai and Rai

Fig. 10.7 Aquatic vegetation harvester depositing on levee (Photo from South Florid Water Management District)

(2003) reported biogas production from water hyacinth and its ability to remove metal and toxic substances while growing in water with the substances. Research has been conducted on heavy metal removal by water hyacinth with impact on plant growth (Muramato and Oki 1983). Significant mercury removal from lagoon wastewater was reported (Chigbo et al. 1982). Experimental study showed that water hyacinth can be used for bioremediation of arsenic removing as much as 600 mg ha^{-1} d^{-1} (Alvarado et al. 2008). On the basis of the capacity of the weed to survive and expand, it demands research on beneficial uses.

10.5   Summary

Water hyacinth and other aquatic vegetation range and extent will keep on expanding due to increased chance of transport and land use changes. The cost of control is expensive. The introduction of water hyacinth in developing countries adds additional burden on their limited resources and challenges their food security. As a result, the weed and similar other weeds get opportunity to expand out of control. The case of water hyacinth infestation and expansion in Lake Tana at the source of the main Blue Nile River demonstrates the situation. When the GERD is completed, the reservoir

Fig. 10.8 Loading of harvested wetland vegetation

Fig. 10.9 Herbicide application on cattails (left) in Lake Okeechobee littoral zone in Florida (right) herbicide refilling (*Photo* South Florida Water Management District; Sharfstein and Zhang 2017; Sharfstein et al. 2015)

will be the second largest water body in Ethiopia and is likely to be infested with water hyacinth flowing down from Lake Tana. Organizational structure with qualified personal and aquatic weed management plan is needed to have better chance of controlling aquatic weeds such as the water hyacinth. Lessons and experiences of integrated weed control from South African and South Florida, among others, can be applied with the objective of monitoring and taking timely steps with filling of the GERD reservoir.

References

Alvarado S, Guedez M, Lue-Meru MP, Nelson G, Alvaro A, Jesus AC, Gyla Z (2008) Arsenic removal from water by bioremediation with the aquatic plants water hyacinth Eichhornia crassipes and Lesser Duckweed (Lemna minor). Biores Technol 99(17):8436–8440

Barrett SCH, Forno IW (1982) Style morph distribution in new world populations of Eichhornia crassipes (Mart.) Solms-Laubach (waterhyacinth). Aquatic Bot 13:299–306

Charudattan R (1986) Integrated control of waterhyacinth (Echhornia crassipes) with pathogen insect, and herbicide. Weed Sci 34 (Suppl, 1):26–30

Chigbo FE, Smith RW, Shore FL (1982) Uptake of arsenic, cadmium, lead and mercury from polluted waters by waterhyacinth Eichhornia crassipes. Environ Pollut Ser A, Ecol Biol 27(1):31–36

Cilliers CJ (1991) Biological control of waterhaycinth, Eichhornia crassipes (Pomtederioceae). Agr Ecosyst Environ 37(1–3):207–217

Gopal B (1987) Waterhyacinth. Elsevier Science

Groote HD, Ajuonu O, Attigton S, Djesson R, Neuenschwander P (2003) Economic impact of biological control of waterhyacinth in Southern Benin. Ecol Econ 45(1):105–117

Harley KLS (1999) The role of biological control in the management of water hyacinth, *Eichhornia crassipes*. Biocontrol News and Inf 11(1):11–22

Heard TA, Winterton SL (2000) Interactions between nutrient status and weevil herbivory in the biological control of waterhyacinth. J Appl Ecol 37(1):117–127

Johnson D (2008) Appendix 5–3: STA herbicide application summary for water year 2007. The 2008 South Florida Environmental Report

Julien MH, Harley KLS, Wright AD, Cilliers CI, Hill MP, Center TD, Cordo HA, Confrancesco AF (1996) International co-operation and linkages in the management of waterhyacinth with impasis on biological control. In: Moran and Hoffmann (eds) Proceedings of the international symposium on biological control of weeds, 19–26 January 1996, Stellenbosch, South Africa, University of Cape Town, pp 273–282

Julien MH (2000) Biological control of water hyacinth with anthropods: a review to 2000. In: Julien et al Proceedings of the second meeting of the global working group for the biological and integrated control of water hyacinth (*Eichhornia crassipes*), Beijing, China, 9–12 Oct 2000

Julien MH, Griffiths MW, Wright AD (1999) Biological control of water hyacinth: the *Neochetina bruchi* and *N. eichhorniae*, host ranges, and rearing, releasing and monitoring techniques for biological control of *Eichhornia crassipes*. ACIAR, Canberra

Kibret S (2017) A concerted effect to save Tana. Global coalition for Lake Tana restoration. http://tanacoalition.org/2017/11/

Lugo A, Bravo-Inclan LA, Alcocer J, Gaytan ML, Olivia MG, Sanchez MR, Vilaclara G (1998) Effect on the planktonic community of the chemical program used to control water hyacinth (*Eichhornia crassipes*) in Guadalupe Dam, Mexico. Aquat Ecosyst Health and Manag 1(1998):333–343

Malik A (2007) Environmental challenges vis avis opportunity: the case of water hyacinth. Environ Int 33(1):122–126

Marlin D, Hill MP, Ripley PS, Aj Strauss, Byrne MJ (2013a) The effect of herbivory by the mite *Orthogalmuna terebrantis* on the growth and photosynthetic performance of water hyacinth (*Eichhornia crassipes*). Aquat Bot 104:60–69

Marlin D, Hill MP, Byrne MJ (2013b) Interactions within pairs of biological control agents on waterhyacinth, Eichhornia crassipes. Biological Control 67(3):483–490

Mishima D, Kunuki M, Sei K, Soda S, Ike M, Fujita M (2008) Ethanol production from candidate energy crops: water hyacinth (*Eichhornia crassipes*) and water lettuce (*Pistia stratiotes* L). Biores Technol 99(7):2495–2500

Muramato S, Oki Y (1983) Removal of some heavy metalsfrom polluted water by water hyacinth (*Eichhornia crassipes*). Bull Environ Contam Toxico 30:170–177

Nigam JP (2002) Bioconversion of water-hyacinth (*Eichhornia crassipes*) hemicellulose acid hydrolysale to motor fuel ethanol by xylose-fermenting yeast. J Biotechnol 97(2):107–116

Nile Basin Discourse. Water hyacinth: a blessing or curse. https://www.nilebasindiscourse.org. Accessed 1 Dec 2017

Opande GO, Onyango JC, Wagai SO (2004) Lake Victoria: the water hyacinth (*Echhornia crassipes* [Mart.] Solms), its socio-economic effects, control measures and resurgence in the Winam gulf. Limnologica 34:105–109

Rodgers L et al (2017) Chapter 7: status of nonindigenous species. The 2017 South Florida Environmental Report

Rodgers L et al (2018) Chapter 7: Status of nonindigenous species. The 2018 South Florida Environmental Report

Sharfstein B, Zhang J, Bertolotti L (2015) Chapter 8: lake Okeechobee watershed annual update. The 2015 South Florida Environmental Report

Sharfstein B, Zhang J (2017) Lake Okeechobee watershed research and water quality monitoring results and activities. The 2017 South Florida Environmental Report

Singhai V, Rai JPN (2003) Biogass production from water hyacinth and channel grass used for phytoremediation of industrial effluent. Biores Technol 86(3):221–225

South Florida Water Management District (2008) Vegetation management strategies for the stormwater treatment areas. West Palm Beach, FL

Tipping PW, Center TD, Sosa AJ, Dray FA (2011) Host specificity assessment and potential impact of *Megamelus scutellaris* (Hemiptera: Delphacidae) on water hyacinth *Eichhornia crassipes* (Pontederiales: Pontededraiceae). Biocontrol Sci Technol 21(1):75–87

Tipping PW, Gettys LA, Meentir CR, Foley JR (2017) Herbivory by biological control agents improves herbicidal control of water hyacinth (*Eichhornia crassipes*). Invasive Plant Science and Manage 10(3):271–276

Van Wyk E, van Wilgen BW (2002) The cost of water hyacinth control in South Africa (2002) a case study of three options. Afr J Aquat Sci 27(2):141–149

Williams AE (2005) Water hyacinth—the world's most problematic weed. Water Encyclopaedia 3:479–484

Wu J, Fu Z, Zhu L (2007) Waterhyacinth in China: a sustainability science-based management framework. Environ Manag 40:823. https://doi.org/10-1007/s00267.007-9003-4

Chapter 11
Financing the Grand Ethiopian Renaissance Dam

Abstract The Grand Ethiopian Renaissance Dam (GERD) is estimated to cost close to 5 billion US dollars, about 7% of the of the 2016 Ethiopian gross national product. The lack of international finance for projects on the Blue Nile River have had long been alleged to Egypt's persistent campaign to maintain presumed hegemony on the Nile water share. Ethiopia is forced to finance the GERD with crowd funding through internal fund raising in the form of selling bond and persuading employees to contribute a portion of their incomes. Despite the domestic success in fund raising, the contribution of Ethiopians and Ethiopian descent living abroad was met with scepticism due to the political environment in Ethiopia. The Chinese government is providing significant amount of international finance to the hydropower infrastructure. Gulf States were also claimed to contribute to the construction of the Dam along the Middle East political line with respect to their relationship with Egypt. Opposition to the government is intertwined with opposition to the fund raising for the dam. The dam is completed over 70% by the end of 2017 and strain of cash is being reported. The successful completion of GERD without explicit support from western financial institutions will have a significant impact on the perception and awareness of Nile water development. The parallels between the planning, construction and financing of High Aswan dam and GERD are stark reminders of critical role of international community to promote co-operation and avoid unintended and lasting ripples on the socio-economic and political landscape of the region.

Keywords Grand Ethiopian Renaissance Dam · Ethiopia · Egypt · Sudan Transboundary rivers · The Nile · GERD financing

11.1 Introduction

Financing major projects in Africa with national fund has been challenging due to Illicit Fund Transfer (IFT) which is estimated between 50 and 148 billion dollars each year (Ighobor and Bafana 2014) citing Economic Commission for Africa. The Nile Basin is shared by eleven countries (Fig. 11.1). The common funding mechanism

© Springer International Publishing AG, part of Springer Nature 2019   161
W. Abtew and S. B. Dessu, *The Grand Ethiopian Renaissance Dam on the Blue Nile*, Springer Geography, https://doi.org/10.1007/978-3-319-97094-3_11

for major projects is foreign aid and loan from international financial institutions such as the World Bank. Since the end of the cold war, western financial institutions has the power to make or break a mega-project. Egypt's diplomatic upper hand has been successful in blocking projects on the Blue Nile for decades. Continuous campaign to persuade international creditors not to finance Ethiopia's Nile basin projects has been reported (Degefu 2003). The recent rise and global interest of china has provided alternative funding source to execute such projects. Egypt lost its leverage on controlling developmental projects in the Nile basin as a result of such alternative fund sources (IDS 2013). The Tekeze dam in northern Ethiopia, on a tributary of the Nile, was financed by the Chinese (Swain 2014). The self-reliance in funding the GERD was the only option with supplemental fund from the Chinese whose international policy is different from the western world.

The Egyptians approach to finance the High Aswan Dam (HAD) provides a valuable insight to the current political and financial challenges and opportunities of the GERD. The feasibility study of both HAD and GERD were conducted by the US. The design and construction of HAD was a pivotal moment in the Egypt-US diplomatic relationship. The dam was also at the center of the cold war where the US and British were determined to prevent the Soviet Union from stepping into Egypt. The US, British and World Bank offered financial support for the construction of HAD as a token for a warm diplomatic relationship with Egypt after World War II. During the planning of HAD, Sudan was under British control and Egypt was contemplating to unify Sudan as its territory. The US and World Bank were concerned about the allocations of water between Egypt and Sudan, that was perceived as a national security threat by Egypt since the agreement would technically tie the water right of Egypt with Britain.

The diplomatic fallout between Egypt and its western allies brought the infamous Swiz-canal crisis in 1956. Egypt nationalized the Swiz-canal in an attempt to use revenues from the canal to finance the HAD. The move triggered military response from British, France and Israel (http://adst.org/2016/06/dont-give-dam-feud-financing-aswan-high-dam/#.WloDNzdG200). Afterwards, Egypt took advantage of the cold war and sought assistance from Soviet Union. The Soviets finally took over the construction of HAD.

The financial support of Western nations and World Bank has the western interests, but more importantly comes with a set of strings mostly to the benefit of Egypt to maintain control over the Nile. Since the fall of Soviet Union, planning and constructions of large infrastructure projects were tied to the blessing of Western powers and the World Bank. However, the international power dynamics has been changing due to the rise of China to some level of the involvement the Soviet Union had in the construction of HAD in the 1950s by providing an alternative funding source for specific functional components of GERD.

The role of Sudan is significantly elevated due to the relative advantage of GERD to the thriving agriculture of Sudan compared to the HAD. This may have played to the irritation of Egypt not to win the support of Sudan on its international diplomatic effort to deter international financing of the GERD.

Fig. 11.1 The Nile River basin and location of GERD

The progress of GERD construction has been reported by domestic and international media. On its seventh year (2011–2018), it is reported that 70% of the dam work is completed (FANA, 4 January 2018), Figs. 11.2 and 11.3. Stress in funding

Fig. 11.2 The status of roller compacted concrete gravity dam, GERD

has been reported and Egyptian papers published Ethiopian government is soliciting fund from Qatar (Middle East Monitor 24 November 2017). Currently Qatar is not in good terms with Egypt and other Arab countries. The meeting of the Ethiopian prime minister and defence minister with Qatar Emir and Qatar defence minister; and Qatar's finance minister visit to the Sudan on November 13, 2017 was reported in relation to GERD finance and politics (Egypt Today 14 November 2017). The GERD project could be dragged into the Middle East politics and conflicts.

Financing GERD would have presented an opportunity to western powers and World Bank towards a peaceful mediation of water right issues of current times and mitigating water conflict. The HAD has been portrayed as a symbol of Egypt's hegemony on the Nile waters, and one may wonder to what effect GERD would change the statuesque.

11.2 Internal Fund Sources

11.2.1 Ethiopian Bank Loans

One of the sources of fund from internal source is Ethiopian banks. The banks have to give low interest loan to the government equivalent to 27% of the loan they give

Fig. 11.3 Concrete face rockfill dam, GERD

out to individuals and businesses (Reuters 23 April 2014). The paper Mentions that such a practice will slow economic growth according to IMF.

11.2.2 Bonds

A bond is a method of raising capital through a promise of a return with relatively longer holding time. The recovery of investment by the bond holder is subject to the credit worthiness and credit maintenance of the bond issuer. The development bank of Ethiopia presents the details of the GERD bond which is issued to Ethiopian nationals and foreigners of Ethiopian origin (dbe.com.net Accessed 15 December 2017). The bond is offered in denominations of Ethiopian Birr 25, 50, 100, 300, 400, 500, 600, 700, 800, 900, 1000, 2000, 3000, 4000, 5000, 10,000, 50,000, 100,000, 200,000, 500,000 and 1 million with no limit. These denominations range in value from $ 0.9 to $ 36,000 in 2017 US dollars. The interest to be earned is 5.5% for maturity up to 5 years and 6% for longer periods. The bond is issued by the Development Bank of Ethiopia, other commissioned agents and regions micro finances. A GERD cup is carried around region to region; nationalistic billboards and related musical performances are used to further mobilize fund collection for the dam out of nationalism. Table 11.1 depicts sources and amount of fund collected from domestic bond sell by 2016. Figure 11.4 chartered cities (Addis Ababa and Dire Dawa).

Table 11.1 GERD bond sell by category (Tesfaye 2016; original source GERD Council, April 2016)

Source category	Bond purchase ($ US, 2017)	Percent
Public and private employees	144,000,000	47
Businesses	57,600,000	19
Diaspora	21,600,000	7
Farmers	28,800,000	9
Others	54,000,000	18
Total	306,000,000	100

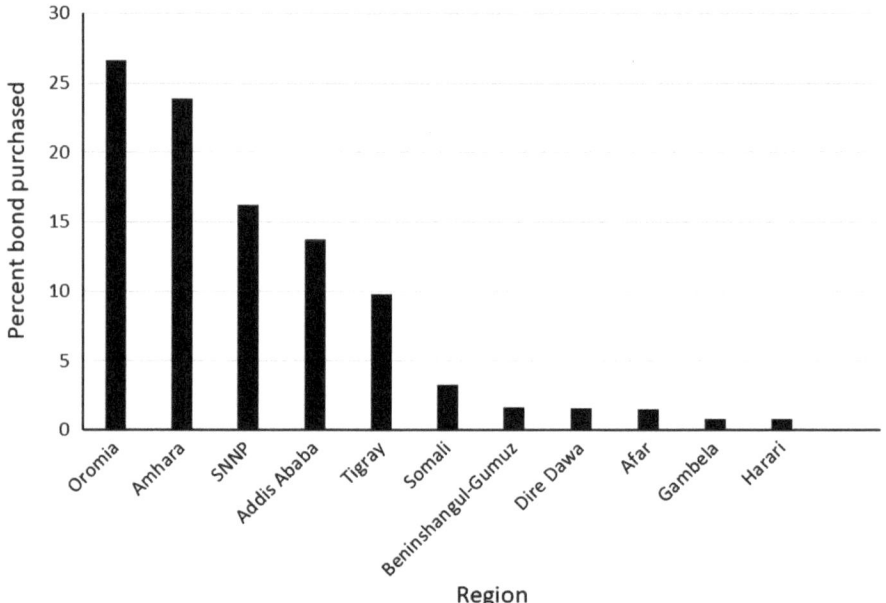

Fig. 11.4 Bond purchase by ethnic region in percent of total (*Data source* Tesfaye 2016; original source GERD Council, April 2016)

11.2.3 Lottery and Other Fund Raising

A mobile phone texting based lottery system with rewards as houses, cars and large sums of money contributes to the capital that is needed to construct the dam. Athletic events are also used to collect fund from the general public.

11.2.4   Civil Servants Salaries

Civil servants have been contributing portion of their salaries for the dam. Opposition parties allege coercion of the program where non participation could affect employee's future career.

11.3   Chinese Fund

The rise of China's international power and capital has made it easier to break into African and other developing countries natural resources, market and strategic partnership. China has become the alternative fund source for projects as the loan is promoted as non-conditional on any internal issues such as mal-governance, human rights and corruption. The high voltage transmission line needed for GERD power transmission is being built by a Chinese firm, State Grid of China Electric Power Equipment and Technology Co. (SGCC), at a cost of 1 billion dollars and the project is completed (Xinhuanet 30 August 2017).

11.4   The Ethiopian Diaspora

The Ethiopian Diaspora in North America, Europe, Australia, the Middle East and other places has a great potential to contribute to Ethiopia's development through contribution of management and technical skills and finance. According to Beyene (2017), failure to raise sufficient fund from the diaspora is attributed to failure to sell the idea of GERD and the bond to the diaspora; and the campaign against the dam by the opposition stating the dam is not a priority and the bond is not a wise investment.

Several attempts to sell bond to the diaspora in several cities raised opposition actions usually disrupting the events and frustrating the effort. In the United States, 5.8 million dollars bond sell to 3100 individuals resulted in a lawsuit on the legality of the bond sell. The United States Securities Exchange Commission found the practice was illegal and fined the Ethiopian Electric Corporation, 6.5 million dollars (https://www.sec.gov/news/pressrelease/2016-113.html). Since there is a potential fund and technical skill among the Ethiopian diaspora, the government of Ethiopia needs to address political concerns to earn the support and participation of an indispensable category. The diaspora's support would have extended to diplomatic support for Ethiopia's water rights. By 2016, only close to 38 million US dollars' worth of bond was sold to the diaspora, mostly in the Middle East (40%), Europe, Africa and North America about 20% each, from close to 3 million diaspora (Tesfaye 2016).

The World Bank which is not funding dams on the Blue Nile in Ethiopia, is said to be funding a 500 kV, 2000-MW high voltage transmission line, Eastern Electric Highway Project, from Ethiopia to Kenya (Kimagai 2016).

11.5 Summary

The financing of GERD transcends international and domestic political and ideological landscape. The GERD presented opportunities for international community to build mutual trust and co-operation towards regional stability and economic development. The international community has made multiple attempts to foster dialogue by organizing regional institutions such as the Nile Basin Initiative. However, the slow progress and lack of autonomy of such institutions is being considered as coerced deterrent from development of physical water resource infrastructure. The planning and construction of GERD mostly reflected the failure of regional co-operation and the historical diplomatic and financial challenges faced by Egypt's High Aswan Dam (HAD) half a century ago. However, the overall process seems to show little change on the effort of major international players to avert destructive conflicts and promote understanding and co-operation between Nile riparian countries. The geo-political strategic importance and economic status of Egypt has been a major factor to successfully fend-off any attempt to acquire funding for physical infrastructure development on the Blue Nile from western financial institutions.

More importantly, the international community has failed to acknowledge the impact of domestic crowed funding and the role of Chinese finance. It should be noted that direct financial contribution of citizens translates to patriotism and political currency to the ruling party in Ethiopia. Hence, the Ethiopian government is less likely to slow down its effort towards completing the GERD unless it is weakened by internal political conflict.

As in the case of HAD, once the GERD is completed, the leverage on financial contribution will not be available and all parties will be compelled to adjust to the reality of the dam. Hence, it is to the best advantage of all stakeholders in the Nile waters and regional politics to weigh in the benefits of the GERD against water conflicts and seek for compromised alternatives. Financial contribution of the downstream countries towards the success of GERD may go a long way fostering sustainable regional development and prosperity.

References

Beyene BM (2017) The grand Ethiopian renaissance dam and the Ethiopian diaspora. http://aigafo rum.com/articles/GERD-and-the-Ethiopian-Diaspora.pdf. Accessed 16 Dec 2017
Degefu GT (2003) The Nile historical legal and developmental perspective. Trafford, Victoria, Canada
http://duwaterlawreview.com/tag/aswan-dam/
Ighobor K, Bafana B (2014) Financing Africa's massive projects. Africa Renewal Online, Dec 2014
Institute of Development Studies (2013) Churning waters: strategic shifts in the Nile basin. Institute of Development Studies. UKAID, Issue 4, Aug 2013
Kimagai J (2016) The grand Ethiopian renaissance dam gets set to open. IEEE spectrum. https://s pectrum.ieee.org/energy/policy/the-grand-ethiopian-renaissance-dam-gets-set-to-open

Swain A (2014) The grand Ethiopian renaissance dam: evaluating its sustainability standard and geopolitics. Energ Dev Front 3(1):11–19

Tesfaye M (2016) The imperative of stepping-up Ethiopian diaspora's contribution to Hedase Dam (GERD). Aigaforum.com. http://aigaforum.com/article2016/GERD-and-Diaspora-041516.pdf. Accessed 17 Dec 2017

Index

© Springer International Publishing AG, part of Springer Nature 2019 171
W. Abtew and S. B. Dessu, *The Grand Ethiopian Renaissance Dam on the Blue Nile*, Springer Geography, https://doi.org/10.1007/978-3-319-97094-3